Forest Management Solutions for Mitigating Climate Change in the United States

Robert W. Malmsheimer, Patrick Heffernan,
Steve Brink, Douglas Crandall, Fred Deneke,
Christopher Galik, Edmund Gee, John A. Helms,
Nathan McClure, Michael Mortimer, Steve Ruddell,
Matthew Smith, and John Stewart

Society of
American Foresters

Society of American Foresters
5400 Grosvenor Lane
Bethesda, Maryland 20814-2198
301/ 897-8720
www.safnet.org

First edition

The Society of American Foresters (SAF) is the national scientific and educational organization representing the forestry profession in the United States. Founded in 1900 by Gifford Pinchot, it is the largest professional society for foresters in the world. The mission of the Society of American Foresters is to advance the science, education, technology, and practice of forestry; to enhance the competency of its members; to establish professional excellence; and, to use the knowledge, skills, and conservation ethic of the profession to ensure the continued health and use of forest ecosystems and the present and future availability of forest resources to benefit society. SAF is a nonprofit organization meeting the requirements of 501(c)(3). SAF members include natural resource professionals in public and private settings, researchers, CEOs, administrators, educators, and students.

Interpretations and conclusions in this publication are those of the authors.

Cite this document as: Malmsheimer, R.W., P. Heffernan, S. Brink, D. Crandall, F. Deneke, C. Galik, E. Gee, J.A. Helms, N. McClure, M. Mortimer, S. Ruddell, M. Smith, and J. Stewart. 2009. *Forest Management Solutions for Mitigating Climate Change in the United States.* Washington, DC: Society of American Foresters.

This report was originally published in the April/May 2008 issue of the *Journal of Forestry* (Malmsheimer et al. 2008).

Library of Congress Cataloging-in-Publication Data
Forest management solutions for mitigating climate change in the United States / Robert W. Malmsheimer... [et al.].
p. cm.
Includes bibliographical references and index.
ISBN 978-0-939970-96-4
1. Forest management--United States. 2. Forest microclimatology--United States. 3. Climatic changes--Environmental aspects--United States. 4. Greenhouse effect, Atmospheric--United States.. I. Malmsheimer, Robert W. (Robert William), 1963-
SD435.F67 2008
581.7'322--dc22

2009004563

Contents

Executive Summary

Forests are shaped by climate. Along with soils, aspect, inclination, and elevation, climate determines what will grow where and how well. Changes in temperature and precipitation regimes therefore have the potential to dramatically affect forests nationwide. Climate is also shaped by forests. Eleven of the past 12 years rank among the 12 warmest in the instrumental record of global surface temperature since 1850. The changes in temperature have been associated with increasing concentrations of atmospheric carbon dioxide (CO_2) and other greenhouse gases (GHGs) in the atmosphere.

Of the many ways to reduce GHG emissions and atmospheric concentrations, the most familiar are increasing energy efficiency and conservation and using cleaner, alternative energy sources. Less familiar yet equally essential is using forests to address climate change. Unique among all possible remedies, forests can both prevent and reduce GHG emissions while simultaneously providing essential environmental and social benefits, including clean water, wildlife habitat, recreation, forest products, and other values and uses.

Climate change will affect forest ecology in myriad ways, with consequences for the ability of forests, in turn, to mitigate global warming. This report summarizes mitigating options involving US forests and examines policies relating to forests' role in climate change. It also recommends measures to guide effective climate change mitigation through forests and forest management, carbon-trading markets, and biobased renewable energy.

Preventing GHG Emissions

Forests and forest products can *prevent* GHG emissions through wood substitution, biomass substitution, modification of wildfire behavior, and avoided land-use change.

Wood Substitution. Substituting wood for fossil fuel–intensive products addresses climate change in several ways. Wood products from sustainably managed forests can be replenished continually, providing a dependable supply of both trees and wood products while supporting other ecological services, such as clean water, clean air, wildlife habitat, and recreation. The use of wood products also avoids the emissions from the substituted products, and the forest carbon remains in storage.

Life-cycle inventory analyses reveal that the lumber, wood panels, and other forest products used in construction store more carbon, emit less GHGs, and use less fossil energy than steel, concrete, brick, or vinyl, whose manufacture is energy intensive and produces substantial emissions.

Although wood product substitution does not permanently eliminate carbon from the atmosphere, it does sequester carbon for the life of the product. Landfill management can further delay the conversion of wood to GHG emissions, or the discarded wood can be used for power generation (offsetting generation by fossil fuel–fired power plants) or recycled into other potentially long-lived wood products. Regardless of the particular pathway followed after a product's useful life, wood substitution is a viable technique to immediately address climate by preventing GHG emissions.

Biomass Substitution. The use of wood to produce energy opens two opportunities to reduce GHG emissions. One involves using harvest residue for electrical power generation, rather than allowing it to accumulate and decay on site or removing it by open field burning. The other is the substitution of woody biomass for fossil fuels.

The use of biomass fuels and bio-based products can reduce oil and gas imports and improve environmental quality. Biomass can offset fossil fuels such as coal, natural gas, gasoline, diesel oil, and fuel oil. At the same time, its use can enhance domestic economic development by supporting rural economies and fostering new industries making bio-based products.

The technologies for converting woody biomass to energy include direct burning, hydrolysis and fermentation, pyrolysis, gasification, charcoal, and pellets and briquettes. Energy uses for wood include thermal energy for steam, heating, and cooling; electrical generation and cogeneration; and transportation fuels.

The United States may need to build 1,200 new 300-megawatt power plants during the next 25 years to meet projected demand for electricity, and coal will likely continue to be a major source of energy for electricity production. Although some energy needs can be met by solar and wind, woody biomass presents a viable short- and mid-term solution: it can be mixed with coal or added to oil- and gas-generated electric production processes to reduce GHG emissions.

Federal funds and venture capital are beginning to support the production of cellulosic ethanol. Substituting cellulosic biomass for fossil fuels greatly reduces GHG emissions: for every BTU of gasoline that is replaced by cellulosic ethanol, total life cycle GHG emissions (CO_2, methane, and nitrous oxide) are reduced by 90.9 percent. The woody biomass is available from several sources: logging and other residues, treatments to reduce fuel buildup in fire-prone forests, fuelwood, forest products industry wastes, and urban wood residues. Plantations of short-rotation, rapid-growing species, such as alder, cottonwood, hybrid poplar, sweetgum, sycamore, willow, and pine, are another source.

Wildfire Behavior Modification. Reducing wildland fires, a major source of GHG emissions, prevents the release of carbon stored in the forest. One modest wildfire—the July 2007 Angora wildfire in South Lake Tahoe, on 3,100 acres of forestland—released an estimated 141,000 tonnes of carbon dioxide and other GHGs into the atmosphere, and the decay of the trees killed by the fire could bring total emissions to 518,000 tonnes. This is equivalent to the GHG emissions generated annually by 105,500 cars.

In 2006, wildfires burned nearly 10 million acres in the United States, and virtually all climate change models forecast an increase in wildfire activity. Under extreme fire behavior scenarios, which could be exacerbated by climate change, increased accumulations of hazardous forest fuels will cause ever-larger wildfires. The proximity of population centers to wildlands significantly increases the risk and consequences of wildfire, including the release of GHGs. Wildfires in the United States and in many other parts of the world have been increasing in size and severity, and thus future wildfire emissions are likely to exceed current levels. Three strategies to reduce wildfires and theirGHGemissions can address that trend:

- pretreatment of fuel reduction areas—that is, removing some biomass before using prescribed fire;
- smoke management—that is, adjusting the seasonal and daily timing of burns and using relative low-severity prescribed fires to reduce fuel consumption; and
- harvesting small woody biomass for energy, or removing some larger woody material (over 10 centimeters, or 4 inches, in diameter) for traditional forest products and burning residuals.

Active forest and wildland fire management strategies can dramatically reduce CO_2 emissions while also conserving wildlife habitat, preserving recreational, scenic, and wood product values, and reducing the threat of wildfires to communities and critical infrastructure.

Avoided Land-Use Change. More carbon is stored in forests than in agricultural or developed land. Preventing land-use change from forests to nonforest uses is thus another way to reduce GHGs. Globally, forestland conversions released an estimated 136 billion tonnes of carbon, or 33 percent of the total emissions, between 1850 and 1998—more emissions than any other anthropogenic activity besides energy production.

Forest conversion and land development liberate carbon from soil stocks. For example, soil cultivation releases 20 to 30 percent of the carbon stored in soils. Additional emissions occur from the loss of the forest biomass, both above-ground vegetation and tree roots.

In the United States, a major threat to forestland is the rise in land values for low density development. Forestland in the US Southeast, for example, has been appraised for forest use at $415 per acre and for urban use at $36,216. Landowners generally convert forestland to residential and commercial uses to capture increasing land values, but when forests are damaged by wildfire, insects, or other disturbances, selling the land for development rather than investing for long-term reforestation can be attractive. Since climate change may increase the prevalence of such disturbances, forestland conversion may increase in the future.

Moreover, conversion of forests to agricultural lands is likely if energy policies favor corn-based ethanol over cellulose-based ethanol. Tax policies that increase the cost of maintaining forestland also promote conversion, as do the short-term financial objectives of some new forest landowners.

Because it is unlikely that publicly owned forestland will increase, efforts to prevent GHG releases from forestland conversion must focus on privately owned forests. New products, such as cellulosic ethanol and new engineered wood products, may add value to working forests. Sustainable utilization of working forests for a combination of wood products, including bioenergy, can improve forest landowners' returns on their land, bolster interest in forest management, and prevent conversion to other uses. Credits for forest carbon offset projects, if trading markets develop, may provide the additional income to encourage private landowners to retain forests.

Reducing Atmospheric GHGs

Forests can also reduce GHG concentrations by sequestering atmospheric carbon in biomass and soil, and the carbon can remain stored in any wood products made from the harvested trees. Because the area of US forests is so vast—33 percent of the land base—even small increases in carbon sequestration and storage per acre add up to substantial quantities.

Sequestration in Forests. The capacity of stands to sequester carbon is a function of the productivity of the site and the potential size of the various pools—soil, litter, down woody material, standing dead wood, live stems, branches, and foliage. Net rates of CO_2 uptake by broad-leaf trees are commonly greater than those of conifers, but because hardwoods are generally deciduous while conifers are commonly evergreen, the overall capacity for carbon sequestration can be similar. Forests of all ages and types have remarkable capacity to sequester and store carbon. There may be potential to sequester or store additional carbon in complex stand structures with mixed species compositions or several age classes.*

Enhancement of sequestration capacity depends on ensuring full stocking, maintaining health, minimizing soil disturbance, and reducing losses due to tree mortality, wildfires, insect, and disease. Management that controls stand density by prudent tree removal can provide society with renewable products, including lumber, engineered composites, paper, and energy, even as the stand continues to sequester carbon. Above all, enhancing the role of forests in reducing GHGs requires keeping forests as forests, increasing the forestland base through afforestation, and restoring degraded lands.

Two active forest management approaches to addressing climate change are 1) mitigation, in which forests and forest products are used to sequester carbon, provide renewable energy through biomass, and avoid carbon losses; and 2) adaptation, which involves positioning forests to become healthier. Adaptive strategies include increasing resistance to insects, diseases, and wildfires; increasing resilience for recovering after a disturbance; and assisting migration—facilitating the transition to new conditions by introducing better-adapted species, expanding genetic diversity, encouraging species mixtures, and providing refugia. This last kind of intervention is highly controversial, however, because action would be based on projections for which outcomes are highly uncertain.

Traditional silvicultural treatments focused on wood, water, wildlife, and aesthetic values are fully amenable to enhancing carbon sequestration and reducing emissions from forest management. Choices regarding even-aged and uneven-aged regimes, species composition, slash disposal, site

*This clarifies a statement regarding mixed-species stands in Malmsheimer et al. (2008) at page 120.

preparation, thinning, fertilization, and rotation length can all be modified to increase carbon storage and prevent emissions. Because forests are the most efficient land use for carbon uptake and storage, landowners with plantable acres and degraded areas that can be restored to a productive condition have a significant opportunity to sequester carbon.

Storage in Wood Products. Harvesting temporarily reduces carbon storage in the forest by removing organic matter and disturbing the soil, but much of the carbon is stored in forest products. The carbon in lumber and furniture, for example, may not be released for decades; paper products have a shorter life, except when disposed of in a landfill. Storage of carbon in harvested wood products is gaining recognition in domestic climate mitigation programs, though accounting for the carbon through a product's life cycle is problematic.

The climate change benefits of wood products lie in the combination of long term carbon storage with substitution for other materials with higher emissions. Because wood can substitute for fossil fuel-intensive products, the reductions in carbon emissions to the atmosphere are comparatively larger than even the benefit of the carbon stored in wood products. This effect—the displacement of fossil fuel sources—could make wood products the most important carbon pool of all.

Forest Carbon Offset Projects

The role of forests and forest products in preventing and reducing GHGs is beginning to gain recognition in market-based policy instruments for climate change mitigation. Forestry is one category of projects that can create carbon dioxide emission reduction credits for trading to offset emissions from industrial and other polluters. Depending on the program, several project types may be eligible: afforestation, reforestation, forest management to protect or enhance carbon stocks, harvested wood products that store carbon, and forest conservation or protection.

Two types of renewable energy credits are becoming available—for using wood-based building materials instead of concrete, steel, and other nonrenewable building materials; and for using wood-based biofuels, such as wood waste, instead of fossil fuels to generate electric power.

Global carbon markets, however, have not yet fully embraced the potential of forests and forestry to mitigate climate change. The Kyoto Protocol, for example, introduced the concept of trading GHG emissions by sources for

GHG removals by sinks, but it limits the role of forestry to afforestation and reforestation. Phase I of the European Union Emissions Trading Scheme allows global trading in carbon dioxide emission reductions to help EU countries reach their targets, but forestry activities are not eligible.

Domestic efforts to date include two regulated emissions trading programs. The Northeast's Regional Greenhouse Gas Initiative, a cap-and-trade program, limits eligibility to afforestation. The other, the California Climate Action Registry, permits credits for afforestation, managed forests, and forest conservation. Voluntary markets for forest carbon include emissions trading transactions through the Chicago Climate Exchange and over-the-counter transactions.

All credit programs must ensure that the net amount of carbon sequestered is additional to what would have occurred without the project. Methods are still being developed to separate the effects of management action on a forest from those of environmental conditions, and determining the net change in carbon stocks must include not only all management actions, such as harvesting, tree planting, and fertilizing, but also the effects of weather, wildfire, insects, and disease.

A forest project must also demonstrate permanence. Ensuring permanence can be difficult, however, since some sequestered carbon might be released through natural events, such as wildfires and hurricanes. Another issue is leakage—the indirect effects that a project might have in, for example, altering the supply of forest products and consequently the total area of forestland.

The current forest carbon accounting principles were developed before forest carbon offsets were recognized as a way for direct emitters of CO_2 to meet emission reduction targets. As a result, they do not adequately address all aspects of using forests to prevent and reduce GHG emissions. Emerging standards for participation in carbon markets may provide consistent rules that are appropriate for managed forests and promote additional and long-term forest carbon sequestration benefits.

Opportunities and Challenges for Society, Landowners, and Foresters

Seven conclusions are apparent from the analyses presented in this report:

1. The world's forests are critically important in carbon cycling and balancing the atmosphere's carbon dioxide and oxygen stocks.

2. Forests can be net sinks or net sources of carbon, depending on age, health, and occurrence of wildfires and how they are managed.
3. Forest management and use of wood products add substantially to the capacity of forests to mitigate the effects of climate change.
4. Greenhouse gas emissions can be reduced through the substitution of biomass for fossil fuels to produce heat, electricity, and transportation fuels.
5. Avoiding forest conversion prevents the release of GHG emissions, and adding to the forestland base through afforestation and urban forests sequesters carbon.
6. Existing knowledge of forest ecology and sustainable forest management is adequate to enable forest landowners to enhance carbon sequestration if there are incentives to do so and if carbon and carbon management have value that exceeds costs.
7. How global voluntary and mandatory markets develop will play a significant role in establishing the price of carbon dioxide and thus creating the incentives to ensure that forests play a significant role in climate change mitigation.

Given those facts, society's current reluctance to embrace forest conservation and management as part of the climate change solution seems surprising. It is beyond argument that forests play a decisive role in stabilizing the Earth's climate and that prudent management will enhance that role. Forest management can mitigate climate change effects and, in so doing, buy time to resolve the broader question of reducing the nation's dependence on imported fossil fuels.

The challenge is clear, the situation is urgent, and opportunities for the future are great. History has repeatedly demonstrated that the health and welfare of human society are fundamentally dependent on the health and welfare of a nation's forests. Society at large, the US Congress, state legislators, and policy analysts at international, federal, and state levels must not only appreciate this fact but also recognize that the sustainable management of forests can, to a substantial degree, mitigate the dire effects of atmospheric pollution and global climate change. The time to act is now.

Tables & Figures

Abbreviations

BTU	British thermal unit
CCAR	California Climate Action Registry
CCX	Chicago Climate Exchange
CDM	Clean Development Mechanism
CER	certified emission reduction
CFC	chlorofluorocarbon
CH_4	methane
CO	carbon monoxide
CO_2	carbon dioxide
CORRIM	Consortium for Research on Renewable Industrial Materials
ERU	emission reduction unit
EU ETS	European Union Emissions Trading Scheme
FT	Fischer-Tropsch (gasification process)
GHG	greenhouse gas
Gt	gigatonne (1 billion tonnes)
GWP	global warming potential (an estimate of the poundfor- pound potential of a gas to trap as much energy as carbon dioxide)
HCFC	hydrochlorofluorocarbon
IIFC	hydrofluorocarbon
HWP	harvested wood product
IPCC	Intergovernmental Panel on Climate Change
JI	Joint Implementation
Mt	million tonnes
MtC/yr	million tonnes of carbon per year
$MtCO_2$ eq.	million tonnes of carbon dioxide equivalents
MW	megawatt
N_2O	nitrous oxide
NMVOC	nonmethane volatile organic compound; also VOC
NO_x	nitrogen oxides
OSB	oriented-strand board
OTC	over-the-counter market
PFC	perchlorofluorocarbon

ppb	parts per billion
ppm	parts per million
REIT	real estate investment trust
RGGI	Regional Greenhouse Gas Initiative
SAF	Society of American Foresters
SF_6	sulfur hexafluoride
t	tonne, or metric ton (1,000 kilograms, 2,205 pounds, or 1.10231 short tons)
TDR	transfer of development rights
Tg	teragram (1,000,000 metric tonnes)
TIMO	timber investment management organization
ton	short ton (2,000 pounds, or 0.907184 metric tonnes)
UNFCCC	United Nations Framework Convention on Climate Change
VER	voluntary (or verified) emission reduction
VOC	volatile organic compound

Preface

In March 2007, on the advice of the Society of American Foresters' Committee on Forest Policy, the SAF Council created the Climate Change and Carbon Sequestration Task Force. Council charged the task force with evaluating the implications of global climate change on forests and forest management, addressing the role of forestry and forests in climate change, offering recommendations for SAF policy activities, and the following tasks:

- briefly assess and summarize the literature on the global climate change implications for forests and their management;
- briefly assess and summarize climate change mitigating options involving forests, including forests' potential as a carbon sink (with cost comparisons to other methods, if information is available), and domestic and international policies relating to forests' role in climate change; and
- recommend possible policy measures to guide effective climate change mitigation through forests and forest management, addressing existing and potential carbon trading markets, opportunities for renewable energy to contribute to mitigation of greenhouse gas emissions, and strategies to minimize the vulnerability and promote adaptation of forests to impacts from climate change.

Prior to publication, the manuscript of this book was reviewed, in whole or in part, by more than 20 scientists. Members of the task force thank all of the reviewers; their efforts increased the report's accuracy and scope. This report and the task force's other products are the result of hundreds of hours by dedicated SAF volunteers.

The task force's report was originally published in the April/May 2008 issue of the *Journal of Forestry* (Malmsheimer et al. 2008). Unless noted, to correct errors or omissions, this book and the *Journal of Forestry* report are identical.

Introduction

Increasing concentrations of greenhouse gases (GHGs) are affecting Earth's climate. Of the many ways to reduce GHG emissions and their atmospheric concentrations, the most familiar are increasing energy efficiency and conservation and using non-GHG-emitting alternative energy sources. Less familiar yet equally essential is using forests to address climate change. Unique among all possible options for reducing or mitigating GHG emissions, forests can both prevent and reduce GHG emissions while simultaneously providing essential environmental and social benefits, including clean water, wildlife habitat, recreation, forest products, and other values and uses.

One approach to reducing current GHG emissions is utilizing excess forest biomass and wood products manufacturing "waste" as a substitute for fossil fuels like oil and coal. Fossil fuels, the net result of ancient photosynthesis, respiration, and decomposition, are replaced only over very long time scales. Current use greatly outstrips their rate of renewal and represents, essentially, the consumption of a nonrenewable resource. A forest biomass energy strategy, however, can use recently sequestered carbon, which can be readily replaced in real time using appropriate forest management systems.

Such a strategy can exceed "carbon neutrality" if the energy inputs required for management, harvesting, transportation, and biomass conversion and combustion are less than the gains obtained through management and long-term use of wood products. Substantial net reductions in carbon dioxide emissions can also be realized by substituting wood products for other construction materials. Reducing the proportion of steel, concrete, brick, and aluminum in buildings and replacing them with solid and engineered wood products leads to large reductions in fossil fuel energy use while storing organic carbon in the buildings for long periods.

Forest management practices can be designed to capture additional atmospheric carbon dioxide, using the capacity of forests to create carbon-based products in the process of photosynthesis. Management can also help forests adapt to changes in temperature and precipitation and thereby maintain their ability to function as carbon "sinks" and deliver other ecosystem services.

If not well managed, however, forests can become a source of GHG emissions, exacerbating climate change. Wildfire, for example, can release enormous amounts of stored carbon. Wildfire is both a creative and a destructive force, particularly in the western United States, and its elimination from the forest landscape is neither possible nor desirable. But active forest management can reduce the severity and extent of forest fires while furthering forest restoration objectives and enabling affordable, effective, and safe fire suppression. Wildfire policy has remained contentious even as the issues have changed; today, we realize that the thinning of overstocked and fire-prone forests can reduce the large-scale release of GHGs from forest fires and simultaneously capture biomass for use as an alternative fuel.

Deforestation is another source of GHG emissions: land-use change releases large amounts of forest-stored carbon back into the atmosphere. Large-scale forest conversion to agricultural land and other uses is taking place in many parts of the world. In the United States, forestland loss is primarily attributable to exurban development and rural housing growth. Progressive policies to encourage the conservation of remaining forests, both in the United States and worldwide, are needed. In particular, we must establish policies that recognize the relation between climate change and forest ownership change and assist forest landowners with stewardship. This includes policies that encourage private forest landowners to become part of the climate change solution through education and carbon market mechanisms that reward forest conservation and carbon sequestration.

As the human impact on Earth increases, we must better understand our connection to the carbon and other GHG cycles and our effect on climate change. The world's current energy strategy is heavily dependent on the use of ancient organic hydrocarbons, and we continue to borrow our energy future from the past. We know that forests play a central role in regulating the carbon cycle, and with the application of sustainable forest management, more net carbon can be transferred into forest carbon sinks. This knowledge and capacity must be used to ameliorate the effects of climate change while we explore solutions for maintaining all the ecosystem services that forests provide.

Addressing GHG emissions requires a multipronged approach (Socolow and Pacala 2006), and forest management is increasingly relevant to carbon sequestration and political and market-based strategies for addressing imbalances in the carbon cycle. Nevertheless, some of today's many national and state legislative proposals do not include the contributions forest management can make to capturing, reducing, and preventing the release of atmospheric carbon dioxide. Certain proposals, for example, recognize the potential of forests as carbon sinks yet marginalize forest management options by creating economic or policy barriers and fail to reward forest landowners for forest conservation and carbon sequestration.

This book discusses opportunities for managed forests and their products to prevent greenhouse gas emissions and sequester atmospheric carbon, and for natural resource managers to help forests adapt to a changing climate. It considers how the science of forest management and the expertise of professional foresters can provide solutions to climate change. It focuses on US forests and the opportunities for their management; it does not attempt to address international issues in any depth.

It outlines sound, science-informed, and practical approaches to the roles that forest conservation and management can play in reducing the concentration of atmospheric carbon dioxide. These approaches can substantially reduce anthropogenic carbon dioxide emissions from land-use change and reliance on fossil fuels. They can also form the foundation for progressive national and international forestry policies designed to foster the prevention and reduction of GHGs and the actual, measurable sequestration of carbon dioxide from the atmosphere.

Organization of the Book

The plan of the book is as follows. The first two chapters lay the groundwork for the mitigation strategies and summarize the science necessary to understand policy options for climate change and greenhouse gases. Chapter 1, Global Climate Change, describes global temperature trends and greenhouse gases, and the relationship between GHG emissions and the activities and sources that generate them. Chapter 2, Potential Effects of Climate Change on Forests, discusses the interrelationship between forests and climate change and the forest related ecological, social, and economic consequences of a warming climate.

The next four chapters detail four ways that forests, forest products, and forest management can prevent GHG emissions. Chapter 3, Preventing GHG Emissions through Wood Substitution, explains how using wood products instead of fossil fuel–intensive products avoids the emissions associated with the latter while storing forest carbon.

Chapter 4, Preventing GHG Emissions through Biomass Substitution, considers the production and use of renewable biomass fuels and other bio-based products to reduce oil and gas imports while improving environmental quality.

Chapter 5, Preventing GHG Emissions through Wildfire Behavior Modification, estimates the amount of carbon released by wildfire, describes how climate change is making catastrophic wildfires more likely, and looks at forest management treatments that can reduce the risks of wildfire.

Chapter 6, Preventing GHG Emissions through Avoided Land-Use Change, examines recent trends in the conversion of forestland to other uses and describes tools that can help retain forest cover and its potential to mitigate climate change.

Chapter 7, Reducing Atmospheric GHGs through Sequestration, turns to the role of forests and forestry in actually reducing GHG emissions, by capturing and storing atmospheric carbon, both in forest carbon "pools" and in long-lived wood products.

Chapter 8, Markets for Forest Carbon Offset Projects, explains carbon trading, a market-based policy instrument that creates incentives to reduce emissions, and the potential for forestry projects to offset emissions from industrial and other polluters.

Finally, Chapter 9, Opportunities and Challenges for Society, Landowners, and Foresters, summarizes the findings presented in the book and stresses the urgency of embracing forest conservation and sustainable forest management as part of the climate change solution.

The forest management solutions for mitigating climate change discussed in this book represent knowledge about forests and climate change as of December 2007 and have undergone scientific review by more than 20 reviewers from a diversity of backgrounds. Science in this area is extensive and evolving, and although our knowledge of forest management solutions to mitigate climate change has greatly improved, contradictions and gaps remain. This book points out contradictions, but since others, such as the US Forest Service (Solomon et al. 2008), are cataloguing knowledge gaps and needs, it does not list areas of needed research.

As we continue to learn how climate change is changing forests and how our changing of forests is changing climate itself, foresters, because of their unique technical training and experience, are best suited to explain the role that science-based forest management and intelligent forest policy can have on climate change. As the largest society of professional foresters in the world, the Society of American Foresters will play a crucial role in guiding these efforts.

Policymakers must realize that managed forests provide climate change solutions that can be adopted quickly and begin preventing and reducing GHG emissions today. Immediacy is critical: many of the other solutions to climate change, such as alternative fuels for energy production and transportation, will not be ready for large scale deployment for years, if not decades. Managed forests solutions can and must help reduce GHG emissions and atmospheric GHGs now.

Chapter 1
Global Climate Change

Global temperatures have fluctuated over the past 400,000 years (Figure 1-1) (US EPA 2007b). Nevertheless, Earth is currently warmer than it has been in its recent past. The Intergovernmental Panel on Climate Change (IPCC) found that "eleven of the last twelve years (1995– 2006) rank among the 12 warmest years in the instrumental record of global surface temperature (since 1850)" (Solomon et al. 2007, 5). The National Research Council concluded "with a high level of confidence that global mean surface temperature was higher during the last few decades of the 20th century than during any comparable period during the preceding four centuries" and, with less confidence, that "temperatures at many, but not all, individual locations were higher during the past 25 years than during any period of comparable length since a.d. 900" (NRC 2006, 3).

As Figure 1-1 indicates, changes in Earth's temperature have been associated with atmospheric carbon dioxide levels in the atmosphere. Research indicates that this and other important gases have also increased recently (Solomon et al. 2007). For example, between the preindustrial period (c. 1750) and 2005,

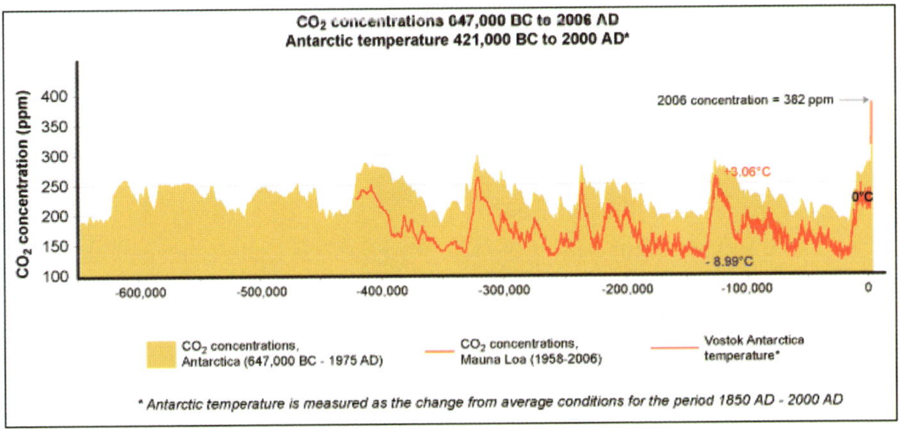

Figure 1-1. Changes in Temperature and Carbon Dioxide (Source: U.S. EPA 2008).

carbon dioxide increased from about 280 parts per million (ppm) to 379 ppm; methane increased from about 715 parts per billion (ppb) to 1,774 ppb; and nitrous oxide increased from about 270 ppb to 319 ppb (Solomon et al. 2007).

IPCC, "the preeminent international body charged with periodically assessing technical knowledge of climate change" (Leggett 2007, 3) and the co-winner of the 2007 Nobel Peace Prize, concluded that "the global increases in carbon dioxide are due primarily to fossil fuel use and land use change, while those of methane and nitrous oxide are primarily due to agriculture," and that these human activities and their byproducts are causing Earth to warm (Solomon et al. 2007, 2). This book does not evaluate the validity of those conclusions, the certainty of the predictions, or whether natural forces are causing changes in the Earth's climate. Rather, our analysis focuses on how climate change may be affecting forests and how managed forests can decrease atmospheric GHG emissions and prevent GHGs from entering the atmosphere.

Greenhouse Gases and the Greenhouse Effect

The biophysical process altering Earth's natural "greenhouse effect" begins when greenhouse gases in the "atmosphere allow the Sun's short wavelength radiation to pass through to the Earth's surface... . Once the radiation is absorbed by the Earth and re-emitted as longer wavelength radiation, GHGs trap the heat in the atmosphere" (Leggett 2007, 22).

Greenhouse gases affected by human activities include carbon dioxide (CO_2), methane (CH_4), nitrous oxide (N_2O), and certain fluorinated compounds—chlorofluorocarbons (CFC), hydrochlorofluorocarbons (HCFC), hydrofluorocarbons (HFC), perchlorofluorocarbons (PFC), and sulfurhexaflouride (SF_6). Other GHGs not directly affected by human activities include water vapor (the most abundant greenhouse gas), carbon monoxide (CO), nitrogen oxides (NO_x), nonmethane volatile organic compounds (NMVOCs, or simply VOCs), and particulate matter or aerosols. NO_x, VOCs, and CH_4 contribute to the formation of another greenhouse gas, ozone (smog), in the troposphere. Most GHGs are generally well mixed around the globe and have global warming effects.

GHGs have different atmospheric lives. For example, water vapor generally lasts a few days, methane lasts approximately 12 years, nitrous oxide 114 years, and sulfur hexafluoride 3,200 years; carbon dioxide's atmospheric life varies (Bjørke and Seki 2005).

GHGs also have different global cycles. For example, the carbon cycle (Figure 1-2) includes geologic, biologic, and atmospheric carbon pools and the cycling that occurs among them (Harmon 2006). Human activities release carbon as carbon dioxide by various methods (described below). These releases alter carbon pools; the most important of these alterations is the transfer of carbon from its geologic pool to its atmospheric pool. Forests play an important role in the carbon cycle because of photosynthesis.

Photosynthesis is the basic process by which plants capture carbon dioxide from the atmosphere and transform it into sugars, plant fiber, and other materials. Within a given land area, this process is known as gross primary production. At the same time, plant respiration, which is necessary for plant growth and metabolism, liberates carbon dioxide back into the atmosphere. The resulting net gain of solid carbon compounds in plant fiber, known as net primary production, can be measured using established forest mensuration

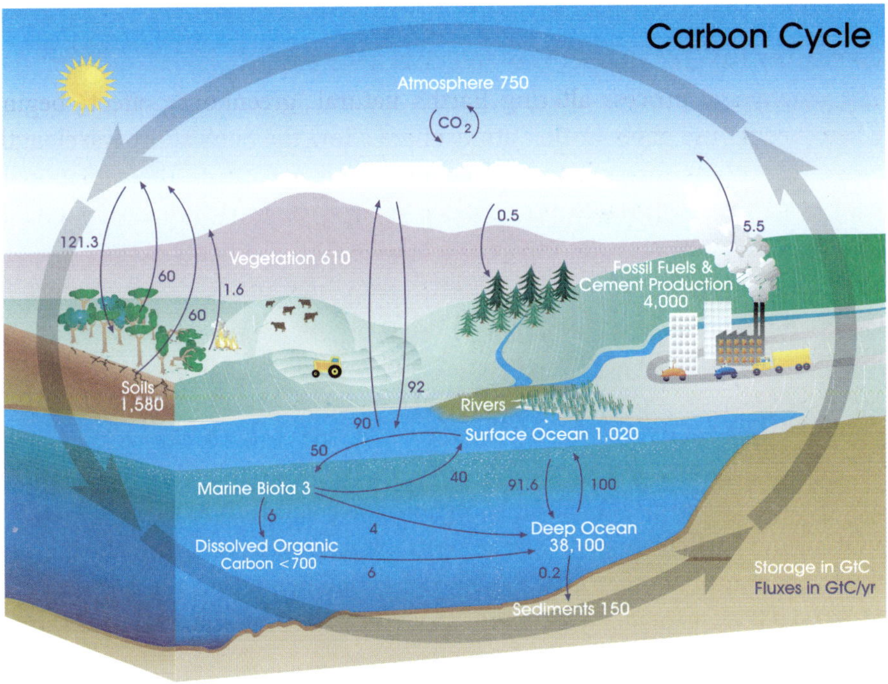

Figure 1-2. Carbon cycle, c. 2004. White* numbers indicate how much carbon is stored in various pools, in billions of tonnes (i.e., gigatonnes, Gt). Purple numbers indicate how much carbon moves between pools each year. The diagram does not include the approximately 70 Gt of carbonate rock and kerogen (oil shale) in sediments (Source: http:// earthobservatory.nasa.gov/Library/CarbonCycle/carbon_cycle4.html).

*This corrects the reference to "black numbers" in Malmsheimer et al. (2008) at page 126.

Table 1-1. Ecosystem Productivity Terms

Term	Definition
Net primary production	Net uptake of carbon by plants in excess of respiratory loss.
Heterotrophic respiration	Respiratory loss by above- and below-ground heterotrophs (herbivores, decomposers).
Net ecosystem production	Net carbon accumulation within the ecosystem after all gains and losses are accounted for, typically measured using ground-based techniques.
Net ecosystem exchange	Net flux of carbon between the land and the atmosphere, typically measured using eddy covariance techniques. The term is equivalent to net ecosystem production but the quantities are not always identical because of measurement and scaling issues.

(Source: Birdsey, U.S. Forest Service, pers. comm., January 2008).

techniques. The overall accumulation of carbon within the ecosystem is known as net ecosystem production (Table 1-1) and includes other net carbon gains, many of which accrue in the soil and are difficult to measure accurately.

Trees and other vegetation store 610 Gt* (610 gigatonnes or 610 billion metric tonnnes) of carbon (Figure 1-2) (1 tonne = 1 metric ton = 1,000 kilograms = 2,205 pounds). In the process of photosynthesis, trees and other plants take CO_2 from the air and in the presence of light, water, and nutrients manufacture carbohydrates that are used for metabolism and growth of both above ground and below-ground organs, such as stems, leaves, and roots. Concurrently with taking in CO_2, trees utilize some carbohydrates and oxygen in metabolism and give off CO_2 in respiration. Vegetation removes a net of 500 million tonnes of carbon dioxide ($MtCO_2$)** (i.e., net primary production) from the atmosphere each year. When vegetation dies, carbon is released to the atmosphere. This can occur quickly (in a fire), slowly (as fallen trees, leaves, and other detritus decompose), or extremely slowly (when carbon is sequestered in forest products). In addition to being sequestered in vegetation, carbon is also sequestered in forest soils. Soil carbon accumulates as dead vegetation is added to the surface or as roots "inject" it into the soil. Soil carbon is slowly released to the atmosphere as the vegetation decomposes (Gorte 2007).

*This corrects the "610,000 tonnes" stated in Malmsheimer et al. (2008) at page 126.

**This corrects the "500 million $MtCO_2$" stated in Malmsheimer et al. (2008) at page 126.

Since GHGs affect the radiative balance of Earth in similar ways, they can be compared using two measures, radiative forcing (externally imposed changes in Earth's radiative balance) or global warming potentials (GWPs); Leggett (2007, 23) calls the latter "an easier but imperfect approximation." GWPs are based on the properties of the most important GHG, carbon dioxide, which is emitted from human sources in by far the greatest quantities (US EPA 2007b). GWPs estimate the pound-for-pound potential of a gas to trap as much energy as carbon dioxide; thus a GWP of 23 indicates that 1 pound of this gas traps as much energy as 23 pounds of carbon dioxide (US EPA 2007b). The global warming potentials of the other principal GHGs are methane, 23; nitrous oxide, 296; hydrofluorocarbons, 120 to 12,000; perfluorocarbons, 5,700 to 11,900; and sulfur hexafluoride, 22,200 (Gerrard 2007).

Greenhouse Gas Emissions

Both natural processes and human activities produce GHGs. Here, drawing on Leggett (2007), we address only the human related sources of the principal GHGs.

- *Carbon dioxide:* combustion of fossil fuels, solid waste, wood, and wood products; manufacture of cement, steel, aluminum, etc.

Table 1-2. Worldwide GHG emissions (CO_2, CH_4, N_2O, PFCs, HFCs, SF_6) by economic sector, 2000.

Sector	$MtCO_2$ eq.	Percentage[a]
Energy	24,722.3	59.4%
Electricity	10,276.9	24.7%
Transportation	4,841.9	11.6%
Manufacturing	4,317.7	10.4%
Other fuel combustion	3,656.5	8.8%
Fugitive emissions[b]	1,629.3	3.9%
Land-use change and deforestation	7,618.6	18.3%
Agriculture	5,603.2	13.5%
Waste	1,465.7	3.5%
Industrial processes	1,406.3	3.4%
International bunker fuels[c]	824.3	2.0%
Total	41,640.5	100.1%

[a] Percentages add up to more than 100 due to rounding.
[b] NO_2 data not available. Fugitive emissions include the leaking of refrigerants from air-conditioning and refrigeration systems.
[c] Fuels used by aircraft and ships.
Source: Data from WRI 2007.

• *Methane:* coal mining, natural gas handling, trash decomposition in landfills, and livestock digestion.

• *Nitrous oxide:* nitrogen fertilizers, industrial manufacturing, and combustion of solid waste and fossil fuels.

• *Hydrofluorocarbons, perfluorocarbons, and sulfur hexafluoride:* commercial, industrial, and household products.

Carbon dioxide is the most prevalent of the GHGs produced by human-related activities. In 2000, it constituted approximately 72 percent of human-related GHG emissions. Methane (adjusted for GWP equivalents) constituted 18 percent, and (adjusted for GWP equivalents) nitrous oxide constituted 9 percent (Leggett 2007). Table 1-2 indicates the human-related activities responsible for the 41,640.5 million tonnes of carbon dioxide equivalents ($MtCO_2$ eq.) of worldwide GHG emissions in 2000 (WRI 2007).

Table 1-3 lists the national shares of the world's GHGs. Relatively few countries produce the most global GHG emissions, in absolute terms, but the "largest GHG emitters have large economies, large populations, or both" (Baumert et al. 2005, 11).

Table 1-3. Ranking of emitters of GHGs (CO_2, CH_4, N_2O, HFCs, PFCs, SF_6), 2000.

Country	$MtCO_2$ eq.	Percentage of world GHGs
1. United States	6,928	20.6%
2. China	4,938	14.7%
3. Russia	1,915	5.7%
4. India	1,884	5.6%
5. Japan	1,317	3.9%
6. Germany	1,009	3.0%
7. Brazil	851	2.5%
8. Canada	680	2.0%
9. United Kingdom	654	1.9%
10. Italy	531	1.6%
Top 10 countries	20,707	61.5%
Rest of world	12,958	38.5%
Developed countries	17,355	52%
Undeveloped countries	16,310	48%

Note: The total world $MtCO_2$ equivalent is different from that in Table 1-2 because Table 1-3 excludes land-use change, deforestation, and international bunker fuels (see Baumert et al. 2005, 12). This table presents the latest available GHG emissions information; countries' current GHG emissions may differ significantly.
Source: Adapted from Baumert et al. 2005, 12.

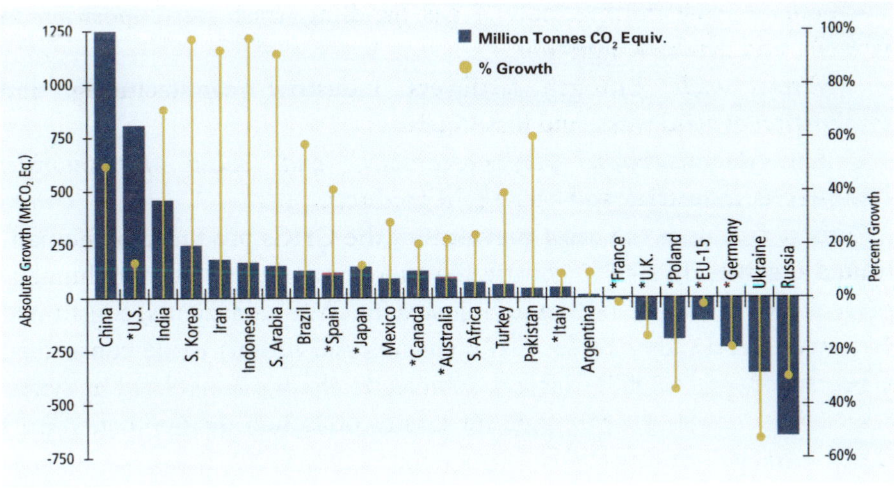

Figure 1-3. Carbon dioxide emissions growth, 1990–2002. * CO_2 plus five other GHGs (Source: Baumert et al. 2005, 15).

Developing countries have the highest emissions growth rates (Figure 1-3). For example, Indonesia's and South Korea's GHG emissions increased 97 percent from 1990 to 2002, Iran's increased 93 percent, and Saudi Arabia's 91 percent (Baumert et al. 2005). China's emissions grew by about 50 percent from 1990 to 2002, but estimates indicate about 35 percent growth for 2003 and 2004 alone (Baumert et al. 2005). Although developed countries' increases are significant in absolute terms, their growth rates are smaller than those of many undeveloped countries.

In 2005, US GHG emissions were 7,260.4* (7,260.4 teragrams, Tg) $MtCO_2$ equivalents (US EPA 2007b). From 1990 to 2005, US emissions rose 16.3 percent as the US gross national domestic product increased by 55 percent (Figure 1-4) (US EPA 2007b). However, because of the sheer size of US emissions, even this relatively small percentage increase in emissions (compared with other countries) contributed considerably to total GHG emissions. For example, USGHG emissions increases from 1990 to 2002 "added roughly the same amount of CO_2 to the atmosphere (863 $MtCO_2$) as the combined 64 percent emissions growth from India, Mexico, and Indonesia (832 $MtCO_2$)" (Baumert et al. 2005, 13).

Future Greenhouse Gas and Global Temperature Estimates

Since "emissions projections require estimating factors such as population, economic growth, and technological change, they are inherently uncertain.

*This corrects the "7,260.4 million" stated in Malmsheimer et al. (2008) at page 127.

... Furthermore, past projections have a weak success record" (Baumert et al. 2005, 18). Nevertheless, all trends point to increasing GHG emissions and global temperatures. For example, the US Energy Information Administration's "midrange" scenario projects that global emissions will rise 57 percent from 2000 to 2025 (Baumert et al. 2005).

The increases are not expected to occur uniformly. For example, China was once expected to surpass the United States as the world's leading GHG emitter in 2020 (Gerrard 2007). However, the country's economic growth has been so fast that the date was moved up to 2009 or 2010. In fact, the most recent reports indicated that China would surpass the United States' CO_2 output by the end of 2007 and that by 2032, "CO_2 emissions . . . from China alone will be double the CO_2 emissions which will come from . . . [the United States,] Canada, Europe, Japan, Australia, and New Zealand [combined]" (Vidal 2007).

IPCC estimates that emissions will result in global warming of about 0.2°C (about 0.36°F) per decade for the next two decades (and even if emissions were held at 2000 levels, a warming of 0.1°C (about 0.18°F) per decade)

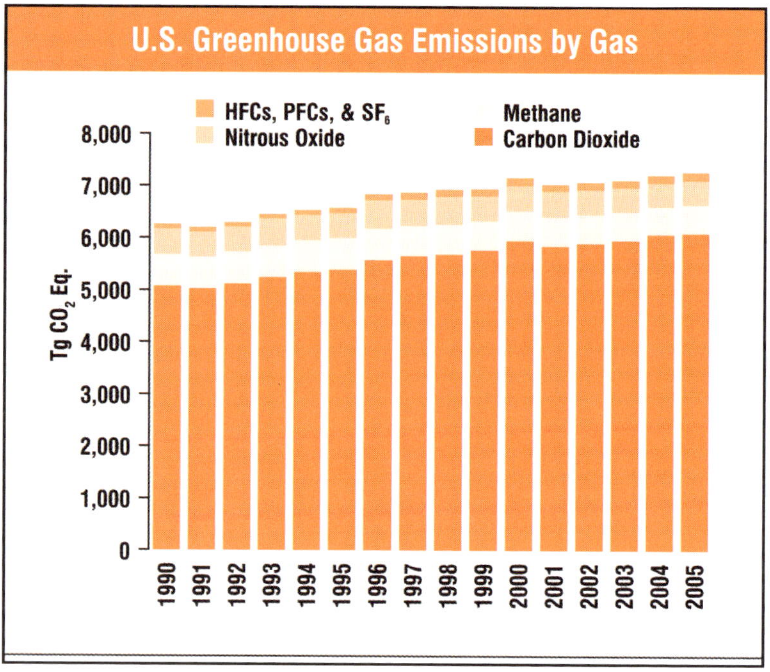

Figure 1-4. US GHG emissions (Source: US EPA 2007b, ES-4).

(Solomon et al. 2007). Longer-term predictions are much less certain, but IPCC scenario projections estimate that global average surface temperature increases (relative to 1980–1999) will range from 1.8° to 4.0°C (3.25° to 7.2°F) for the 2090–2099 decade (Solomon et al. 2007).

Decades after the first generally recognized indications of global warming, the science of climate change remains contentious. While some scientists contend that the Earth's atmosphere is warming, polar ice caps are shrinking, and sea levels are rising because of anthropogenic increases in the concentrations of greenhouse gases, some say that the presumed causes are wrong, the reports overstated, and the predictions mistaken (e.g., Singer 2008; Bast and Taylor 2007; McKitrick et al. 2007). What is not at issue, however, is that forests play a central role in the balance of carbon stocks on Earth, and the policies now being developed and implemented to address climate change will be the more effective the more they incorporate forestry.

Chapter 2
Potential Effects of Climate Change on Forests

Forests are shaped by climate. Along with soils, aspect, inclination, and elevation, climate determines what will grow where and how well. Changes in temperature and precipitation regimes therefore have the potential to dramatically affect forests nationwide.

Climate is also shaped by forests. Forest stands act as windbreaks, and forest canopies influence the interactions of soil, water, and temperature. Forests can act as a carbon sink, helping to offset greenhouse gas emissions; in 2003, US forests sequestered more than 750 million tonnes of CO_2 equivalent (US EPA 2005). Alternatively, afforestation in certain areas may reduce surface reflectivity, or albedo, such that any reductions in radiative forcing (warming) gained from increases in carbon sequestration are offset (Betts 2000). The interrelationship between forests and climate means that dramatic change to one will influence the other. In some situations, this feedback is negative, dampening further iterations. In other situations, however, this feedback is positive, building upon and exacerbating the initial change (e.g., Woodwell et al. 1998; Fleming et al. 2002).

The role of climate as a driver in ecosystem function is well established (e.g., Stenseth et al. 2002). A changing climate will affect forests in several ways, ranging from direct effects of temperature, precipitation, and increased atmospheric concentrations of carbon dioxide on tree growth and water use, to altered fire regimes and changes in the range and severity of pest outbreaks. Climate change has the potential to transform entire forest systems, shifting forest distribution and composition. Economically, climate change is expected to benefit the timber products sector (e.g., Irland et al. 2001). Overall harvests in the United States are expected to increase. In terms of lost timber value, suppression costs, and loss of recreation and ecosystem services, however, the costs of wildfire are expected to increase dramatically. Importantly, the specific implications of climate change for forests will vary greatly from place to place.

Ecological Effects

Global mean surface air temperature is expected to increase over the next century, as described in Chapter 1. Temperature minimums are expected to increase faster than maximums, and the growing season is likely to lengthen, especially in the middle and high latitudes (IPCC 2007). Changes in precipitation are likewise expected: tropical and high-latitude areas may experience increases in precipitation, and the subtropics and middle latitudes are expected to experience decreases (IPCC 2007). Heat waves will likely be greater in terms of frequency, intensity, and duration, while precipitation will become more intense but with longer intervals between events.

Climate change and an increased concentration of atmospheric carbon will affect forests on multiple levels. At the individual tree level, an increase in atmospheric carbon dioxide concentrations is expected to lead to increased levels of net primary productivity and an increase in overall biomass accumulation, primarily in the form of fine root production but potentially also through allocation to woody biomass (Ainsworth and Long 2005; Calfapietra et al. 2003; Norby et al. 2002, 2004, 2005). The exact response to elevated carbon dioxide concentrations, however, may vary by species and locale (Norby et al. 2002; Korner et al. 2005; Handa et al. 2005). In forests where photosynthesis is limited by CO_2 concentrations, the degree to which such an increase can be sustained over time will be limited by other factors, such as the availability of nitrogen or water (Kramer 1981; Norby et al. 1999; J.G. Hamilton et al. 2002). Active fertilization may allow for increased productivity under elevated atmospheric carbon dioxide concentrations, especially on nutrient-poor sites (Oren et al. 2001; Wittig et al. 2005). Apart from effects on individual productivity, increased atmospheric carbon dioxide concentrations are also expected to alter leaf chemical composition, affecting herbivore fitness as a result (Saxe et al. 1998). These latter ramifications have been shown to vary across species and other environmental variables, such as temperature (Lincoln et al. 1993; Bezemer and Jones 1998; Zvereva and Kozlov 2006).

Either in addition to or in concert with increased concentrations of atmospheric carbon dioxide, climate change–induced shifts in temperature and precipitation regimes are expected to affect individual trees' fitness and productivity as well (Saxe et al. 1998; Nabuurs et al. 2002; Sacks et al. 2007). Changes in absolute temperatures (e.g., frost, heat stress) as well as changes in the form, timing, and amount of precipitation (e.g., snow versus rain, drought versus flood) can affect forests directly. In boreal, temperate, and Mediterranean European forests, temperatures are expected to increase along with precipitation, raising productivity (Nabuurs et al. 2002). Other regions may experience increasing temperatures along with a decrease in absolute precipitation or a shift in the form of precipitation, possibly changing the seasonal availability of water in the form of snowpack or snowmelt and causing seasonal water shortages (Barnett et al. 2005; Trenberth et al. 2007). A water shortage can also counteract any productivity benefits from increased atmospheric carbon dioxide concentrations or a longer growing season (Wullschleger et al. 2002). Other atmospheric constituents can further exacerbate temperature and precipitation stressors. In particular, nitrogen deposition rates and ozone concentrations, which are expected to rise (IPCC 2007; Nabuurs et al. 2002), can magnify the effects of drought (Schlyter et al. 2006; Eatough-Jones et al. 2004).

The effects of climate and atmosphere on individual trees are borne out at the stand and forest system levels because individual fitness also influences susceptibility to pests, pathogens, and severe weather events (Schlyter et al. 2006). In addition, a warmer climate will likely allow herbivores and pests to expand in both number and range (Logan et al. 2003). For example, milder winters are expected to decrease winter mortality in white-tailed deer, exacerbating browse and forage damage (Ayers and Lombardero 2000). Species such as the rocky mountain pine beetle and the southern pine beetle are expected to expand their ranges, not only latitudinally but altitudinally as well, possibly exposing jack pine (*Pinus banksiana*) and whitebark pine (*P. albicaulis*) to new or increased levels of attack (Logan and Powell 2001; Williams and Liebhold 2002). In northern Europe, the spruce bark beetle, in the past usually limited to a single brood per season, will likely produce multiple broods with increasing frequency (Schlyter et al. 2006). In all, a warmer climate is expected to encourage pest outbreaks of increasing frequency, duration, and intensity (Volney and Fleming 2000; Logan et al. 2003; Gan 2004).

Climate change is also predicted to alter the frequency and intensity of severe weather events (Opdam and Wascher 2004; IPCC 2007). Any change in frequency or intensity, coupled with a change in individual or stand fitness brought about by changes in temperature, precipitation, or outbreaks of pests or pathogens, will affect forests. Species range and distribution may change as a result (Opdam and Wascher 2004).

Increases in the amount of downed or damaged timber, whether caused by weather, pests, or pathogens, combined with the direct effects of shifting temperature or precipitation patterns will strongly influence fire regimes. The effect may be exacerbated by another driver of fire, the increased human presence in the wildland-urban interface. In some areas, such as the Canadian boreal forest, an increase in precipitation may actually lead to a decrease in fire activity relative to historical rates (Bergeron et al. 2004). But in the western United States, climate change is thought to be a primary driver of the recent increase in fire frequency and duration (Westerling et al. 2006). In extreme cases, climate change–induced increases in fire severity and frequency may even facilitate the conversion of forestland into grassland (Flannigan et al. 1998). Although forest management and fuel removal may help counter the increased severity, intensity, and duration of wildfire, such activities may be insufficient to address the full effects of climate change on fire regimes (Westerling et al. 2006).

Under a changing climate, the combined and individual influences of temperature, precipitation, atmospheric carbon dioxide concentration, pests, weather, and fire will have dramatic effects on forest systems. The consequences will be seen in the distribution and composition of forests across entire landscapes. In particular, forest types are expected to migrate both latitudinally and altitudinally (Walther et al. 2002). In the Rocky Mountain zone, for example, a 3.5°C (6.3°F) increase in temperature is expected to shift habitat more than 2,000 feet in elevation or 200 miles north (Ryan 2000). Past episodes of climate change have witnessed forest migration rates of approximately 50 kilometers per century, with some species achieving even greater rates of migration (Schwartz 1993; Noss 2001). The current rate of climate change may exceed the rate at which forests can respond (Woodwell et al. 1998). To match current rates of warming, northward shifts of 500 kilometers over the next century may be necessary—a migration rate up to one order of magnitude greater than that witnessed in the past (Schwartz 1993). Past, present, and future fragmentation of forestland may inhibit dispersal and establishment, significantly reducing potential migration rates (Schwartz et al. 2001; Walther et al. 2002; Opdam and Wascher 2004). As a result, future rates of migration are expected to be at least one order of magnitude *slower* than those seen in the past (Schwartz 1993). Still, evidence does exist of long-distance migration over relatively short time frames (Clark 1998), and disturbance may actually facilitate dispersal by opening canopy

(Schwartz et al. 2001). However, an increasing frequency of large-scale disturbances is likely to facilitate the spread of invasive species into forest systems as well (Iverson and Prasad 2002).

Adaptation is another mechanism by which forests can respond to climate change, but it is likely to occur at rates well below what is necessary to respond to expected changes (Opdam and Wascher 2004). A failure to adapt or migrate could result in species extirpation or extinction, or the conversion of forest to grassland or other systems (Iverson and Prasad 2002; Woodwell et al. 1998). This can be counteracted, at least to some extent if not entirely, by active forest management, including facilitated dispersal (Schwartz et al. 2001). Even with active forest management, however, species' ranges may shift enough to result in local, regional, or even national extirpation. For example, models have indicated that under various scenarios accompanying a doubling of atmospheric carbon dioxide concentrations, quaking aspen (*Populus tremuloides*), bigtooth aspen (*P. grandidentata*), sugar maple (*Acer saccharum*), paper birch (*Betula papyrifera*), and northern white cedar (*Thuja occidentalis*) all face potential extirpation from the United States (Iverson and Prasad 2002).

Shifts in forest species composition and range, along with the already-mentioned changes in temperature, precipitation, and fire regimes, will likely have tremendous implications for forest biodiversity. Widespread species response to climate change has already been documented (Parmesan and Yohe 2003). Old-growth forests may suffer increased mortality rates (e.g., van Mantgem and Stephenson 2007), with possible implications for wildlife habitat. Warming trends may lead to mismatches in the timing of once-synchronous events, such as bud burst, moth hatching, and peak food demand by nesting birds (Walther et al. 2002). Range shifts by individual species may alter system dynamics, resulting in new relationships and associations (Skinner 2007). Species with narrow niches will likely face decline or loss (Kirschbaum 2000). Changes in the form and amount of precipitation, along with associated water availability within a forest ecosystem, may directly affect bird, amphibian, and reptile communities by concentrating populations and increasing their vulnerability to parasites and pathogens (Pounds et al. 1999). Protected areas, the boundaries of which are largely static, may cease to protect targeted species, processes, features, or attributes (Halpin 1997; Burns et al. 2003). Some US protected areas may lose up to one-fifth of the species currently found within their boundaries, but expanding northern ranges may result in a net increase in the total number of species these areas contain (Burns et al. 2003).

Appropriate forest management can help reduce the negative effects of climate change on forests. A variety of management options and objectives exist, but recent comprehensive reviews suggest that no single management strategy will satisfy all needs in all situations (Millar et al. 2007). Apart from the aforementioned facilitated dispersal and fuels treatment activities, adjustment of rotation lengths and regional harvesting patterns can likewise mitigate the negative effects of climate change (Easterling et al. 2007). Preemptive harvesting of vulnerable stands, for example, may help contain pest outbreaks (Volney and Fleming 2000), and preventing further forest fragmentation and maintaining gene pools can help ensure that forest function and diversity are preserved (Noss 2001).

Social and Economic Effects

Climate change is expected to affect social and economic aspects of forests as well as forest ecology. The implications for nonwood forest products and services, such as biodiversity, recreation, and edible plants, are difficult to assess, in part because of the high uncertainty regarding the ecological effects.

The effects of climate change on one social aspect of forests, forest-based recreation, are complex. Some activities will witness a net benefit while others will suffer, depending on the type of activity, the seasonal nature of the activity, and the incidence of extreme weather events (Irland et al. 2001). Beach recreation at mountain lakes might benefit as a result of extended seasons, for example, but other uses that are sensitive to average temperatures and climatic variability, such as cold water stream fishing and snow skiing, could lose (Alig et al. 2004). Lake-based recreation could be negatively affected if lake levels fall because of increased evapotranspiration and changing precipitation patterns (Irland et al. 2001).

Economic effects are likely to vary regionally, but uncertainty over specific impacts remain. Forest productivity is anticipated to continue to slowly rise as demand for industrial wood demand also climbs modestly. Globally, timber production is expected to increase (Easterling et al. 2007). The United States may see a net benefit to the timber products sector, with saw timber production increasing relative to pulpwood (Irland et al. 2001). Future prices for solid wood and pulp have been examined in at least seven model simulations of climate change (Easterling et al. 2007), and it is expected that supply will meet demand. Most models predict price declines for both solid wood and pulp, which means consumers and mill owners would experience net benefits while landowners and producers would experience net losses (Irland et al. 2001). Harvests are expected to increase across large portions of the United States, especially in the South, where existing infrastructure and lower costs are favorable (Joyce et al. 1995).

Dramatic northern migration of forests accompanied by increasing dryness across the South could result in an opposite outcome, however, shifting increased production to the North (Shugart et al. 2003). Economic effects are predicted to be most sensitive to migration of southern pine northward, which could lead to positive economic outcomes; less optimistic are predictions of no increase in growth or perhaps even a decline in growth for southern softwood timber.

Any economic benefits resulting from changes in mean temperature or precipitation are likely to be outweighed by extreme events of increasing severity and frequency (Easterling et al. 2007). Research indicates that the local economic consequences, positive or negative, of increased extreme weather events can vary. Those events that limit site access may restrict supply in the short term, but events resulting in down or damaged timber and therefore salvage opportunities may increase short-term supply, with different consequences for private landowners and government agencies (DeWalle et al. 2003).

Plant damage from pests, such as the mountain pine beetle in the West and the gypsy moth in the East, will continue to be significant should recent warming trends and drought continue. Although quantitative analyses and modeling of climate change–related pest infestations are somewhat limited, studies have predicted that annual damage from the southern pine beetle alone will increase by four to seven-and-a-half times current levels, or $492 million to $869 million per year (Gan 2004). Furthermore, any pest damage will be amplified by climate extremes. Past research has attempted to capture the effects of interrelated stressors and disturbances (e.g., Fleming et al. 2002), but as yet, few models can fully simulate these interactions (Easterling et al. 2007).

More rain and less snow will mean greatly reduced spring snowpacks; California may see up to a 90 percent reduction in spring snowpack by the end of this century. Smaller snowpacks will lead to longer, drier summers and greatly increase the risk of wildfire and pests. The interactions between pest infestations and wildfire can enhance one another. A recent climate modeling study for the West shows that for Washington State, average annual area burned could expand two to five times by the end of this century (Casola et al. 2006). Modeling in California indicates up to a 55 percent increase in annual average area burned by the end of this century (Cayan et al. 2005). Larger burns would require continued and substantial increases in fire prevention and suppression costs—in just the past 18 years, fire management expenses of the US Forest Service have increased from 13 to 45 percent of the agency's budget (USFS 2007)—and mean an even heavier burden on both federal and state governments. One consequence is corresponding reductions in other resource programs. Moreover, wildfire's economic effects—on timber value, recreation receipts, ecosystem services such as water quality and quantity, and human health related to air pollution—all could generate costs and consequences that are many times larger than the fire prevention, preparedness, and suppression costs (Climate Leadership Initiative 2007). Many other substantial economic costs due to wildfire will be felt across much of the nation. One example involves watershed effects: burned areas produce 25 times more sediment than unburned areas, with profound implications for debris cleanup (Loomis et al. 2003), including dredging of reservoirs.

Perhaps the largest economic effect on forests and forest management would come as a result of a cap on carbon emissions. Carbon pricing, considered an essential element of emissions mitigation policy (Stern et al. 2006), could increase the use of fuelwood or forest biomass relative to traditional fossil fuels (Easterling et al. 2007). A carbon price in conjunction with an established offset market could likewise encourage significant increases in domestic carbon sequestration through afforestation and changes in forest management practices (US EPA 2005).

Forest management can play a critical role in minimizing the negative effects of climate change on forests while maximizing positive ones (Shugart et al. 2003; Easterling et al. 2007). Specialized equipment (e.g., harvesters and trucks that achieve high fuel efficiency and minimize soil displacement) can help offset the negatives (DeWalle et al. 2003). Other mitigation or adaptive actions involve changes in gene management, forest protection, forest regeneration, silvicultural treatment, forest operations, maintenance of nontimber resources, and park management (Spittlehouse and Stewart 2003).

Chapter 3

Preventing GHG Emissions Through Wood Substitution

Wood substitution addresses climate change in several ways. Wood products from sustainably managed forests can be replenished continually, providing a plentiful and dependable supply of both trees and wood products while supporting other ecological services, such as clean water, clean air, wildlife habitat, and recreation (USFS 2005). Substituting wood for fossil fuel–intensive products also avoids the emissions from the substituted products, and what was forest carbon remains stored in the wood products.

Trees remove carbon dioxide (CO_2) from the atmosphere and store it in their roots, stems, trunks, and leaves through the process of photosynthesis. In addition, forested ecosystems store carbon in soil, forest floor, and down dead wood. As forests and their trees mature, their growth slows; however, some studies indicate that as tree growth slows, ecosystem storage of carbon may actually increase as a result of increases in other carbon pools (Zhou et al. 2006; Schulze et al. 2000). Although more definitive research is needed, it appears that both short-rotation management and long-rotation or old-growth management can lead to greater overall carbon sequestration. Intensively managed commercial forests, using short rotations, can sequester significant carbon if the wood products are long-lived (Perez-Garcia et al. 2005). Long rotations and old-growth management mean little or no carbon is stored in wood products but more carbon is stored in the ecosystem. If the only forest management goal is to sequester carbon, both short-rotation intensive management and old-growth management are appropriate; however, if the goal is also to produce wood products, then short rotation management that leads to long lived products would be the approach of choice.

Life-Cycle Assessments

Public interest in the environmental impacts of forest management has created demand for strategies and policies to improve environmental performance, some of which can have unintended consequences. Harvest reductions, for example, alter the availability of wood, and in turn, the price of building materials. This increases wood imports from other countries or causes consumers to use nonwood substitutes. The environmental consequences of these changes in material flow and uses are difficult to quantify because of the complexity of tracking materials through market transactions (USFS 2005), but contrary to intuition, the use of nonwood substitutes is often detrimental to the environment.

What exactly are the environmental benefits of substituting wood for steel and concrete? The Consortium for Research on Renewable Industrial Materials (CORRIM) was created as a not-for-profit consortium by 15 research institutions to update and expand a 1976 report by the National Academy of Sciences on the effects of producing and using renewable materials (Lippke et al. 2004). CORRIM developed a complete life cycle inventory of all environmental inputs and outputs, from forest regeneration through product manufacturing, building construction, use, maintenance, and disposal. It constructed virtual houses (of approximately 2,250 square feet, an average size) and used a life-cycle assessment to determine the associated energy use, air and water pollution, global warming potential, and solid waste production (Lippke et al. 2004). The virtual houses, using framing materials of wood, steel, and concrete, were "built" in two very different locations: Minneapolis (wood versus steel) and Atlanta (wood versus concrete).

Figure 3-1 depicts the life-cycle assessment for a wood-frame house. It includes transportation for each stage from forest management (regeneration) to harvesting, product manufacturing, building construction, use and maintenance, and recycling or disposal. Each stage of processing had different effects, providing insight into where opportunities for improvement could have the greatest overall benefit.

Forest resource management can positively affect climate change. However, implementation of any kind of management treatment requires forest operations, such as harvesting, processing or conversion, and transportation of biomass. These operations affect the GHG profile of forestry activities through the direct emissions of the equipment and the relative efficiency of handling biomass volume (Brinker et al. 2002). Operators employ a wide range of equipment and operational methods, from loggers with chainsaws to highly mechanized mass-production logging systems, to reduce environmental impacts and create economic efficiencies. Power technologies for forest equipment are changing with federally mandated

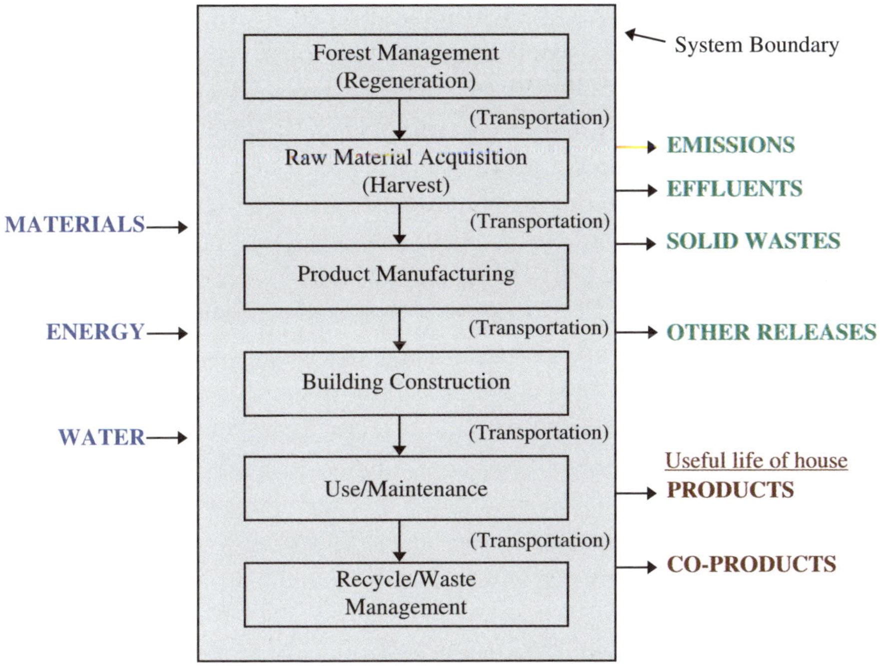

Figure 3-1. Life-cycle assessment from regeneration of trees to disposal of wood materials (Source: CORRIM Presentations, www.corrim.org/ppt/2005/fps_june2005/lippke/index.asp).

transitions to different fuel types and cleaner diesel engines, and alternative-fuel equipment, including hybrids and biofueled machines, is being tested. Emission reductions must be assessed on a net basis. A low emissions system may be relatively inefficient at processing carbon volume and thus a poor choice under climate change scenarios (Brinker et al. 2002). However, the energy requirements for harvesting and transportation are substantially lower than for product manufacture, where the energy required for drying is a major factor but can largely be provided by biofuels with negligible net greenhouse gas emissions (Puettmann and Wilson 2005).

Life-cycle inventory analysis reveals that the wood products used in construction store more carbon and use less fossil energy than steel, concrete, brick, or vinyl. Conversely, the manufacture of nonwood products is energy intensive and produces substantial emissions, including global warming potential (GWP) emissions (K. Skog, US Forest Service, Forest Products Laboratory, pers. comm., November 2007).

Table 3-1 presents the summary environmental performance indices for typical Atlanta and Minneapolis houses built to code. With two exceptions (solid waste in the Minneapolis house and water pollution in the Atlanta house), the index measures for the wood-frame designs are considerably lower than for the nonwood frame designs. Notice that for global warming potential, steel has 26 percent higher CO_2 equivalent than wood, and concrete, 31 percent higher CO_2. The difference is particularly significant considering that the framing accounts for only about 6 percent of the mass of the house; the rest of the house's materials are unchanged.

Life-cycle assessment of building systems, like walls and floors, shows that carbon emissions are very sensitive to design and product selection, with steel and concrete walls and floors producing several times more emissions than wood-dominant assemblies (Lippke and Edmonds 2006). Figure 3-2 shows the GWP differences for four floor designs, not including any insulation or floor covering. The concrete floor produces more than four times the GHG emissions of a dimensional lumber or wood joist floor. The steel design is much worse, releasing 731 percent more GWP than wood I-joist floors, largely because the horizontal application of steel in a floor requires a high gauge to reduce bending and bounce.

Figure 3-3 shows similar comparisons for an Atlanta wall, including insulation and cladding. The increase in GWP for the concrete wall over a

Table 3-1. Environmental performance indices for residential construction.

Minneapolis home	Wood frame	Steel frame	Difference	Percentage Change
Embodied energy (gigajoules)	651	764	113	17
Global warming potential (kg CO_2)	37,047	46,826	9,779	26
Air emissions index (index scale)	8,566	9,729	1,163	14
Water emissions index (index scale)	17	70	53	312
Solid waste (total kg)	13,766	13,641	−125	−0.9
Atlanta home	**Wood frame**	**Concrete frame**	**Difference**	**Percentage Change**
Embodied energy (gigajoules)	398	461	63	16
Global warming potential (kg CO_2)	21,367	28,004	6,637	31
Air emissions index (index scale)	4,893	6,007	1,114	23
Water emissions index (index scale)	7	7	0	—
Solid waste (total kg)	7,442	11,269	3,827	51

Source: Lippke et al. 2004, 13.

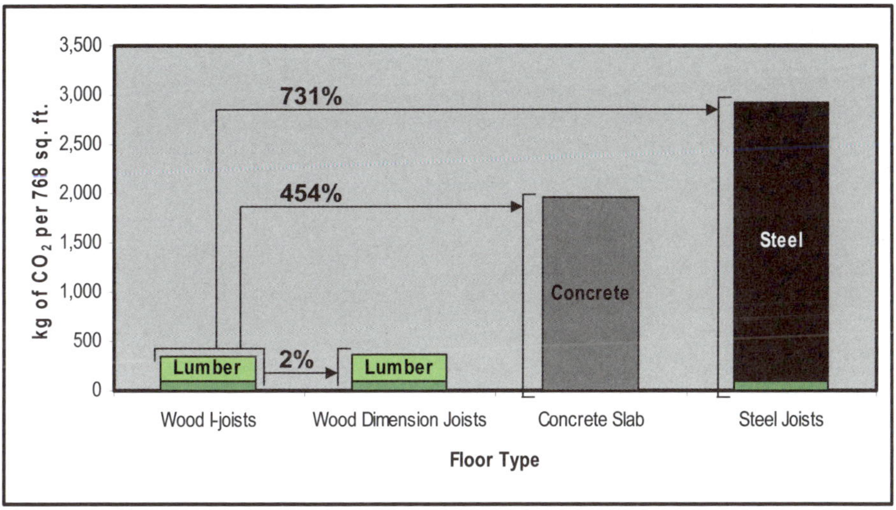

Figure 3-2. Global warming potential of alternative floor materials (Source: Lippke and Edmonds 2006, 63).

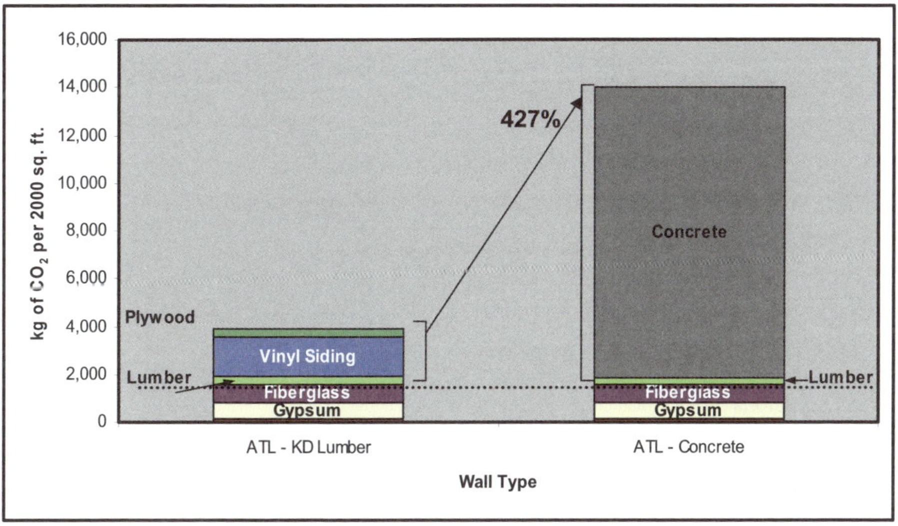

Figure 3-3. Global warming potential of alternative wall-framing materials, Atlanta. Materials below the dotted line are common to both wall designs (Source: Lippke and Edmonds 2006, 61).

kiln-dried lumber wall is similar to the floor comparison. The calcification process used to produce concrete increases the GWP for the concrete design's block, stucco, and lumber frame 427 percent compared with the kiln-dried lumber design's plywood, vinyl, and lumber (Lippke and Edmonds 2006).

Wood use can substantially alter environmental performance and reduce emissions, especially when wood is substituted for fossil fuel–intensive products and energy. For example, for a Minneapolis steel stud wall, the steel and its required insulation have 44 percent higher GWP than the kiln-dried wood wall; both walls' cladding and gypsum contribute almost as much to emissions as the framing elements (Figure 3-4, columns 2 and 3). However, substituting wood siding for vinyl siding, wood paneling for gypsum, cellulose for fiberglass, and increasing biofuel use for drying reduces emissions by 75 percent (Figure 3-4, column 1).

Figure 3-5 illustrates the integrated effect of all carbon pools present in a forest as it matures, along with the carbon removed by product pools based on the life-cycle assessment. It shows a modest increase of carbon in the combined forest and product pools over time (yellow* line), unlike the steady state that exists in a forest (green line; i.e., when wood products are not removed). More importantly, as wood products are substituted for fossil

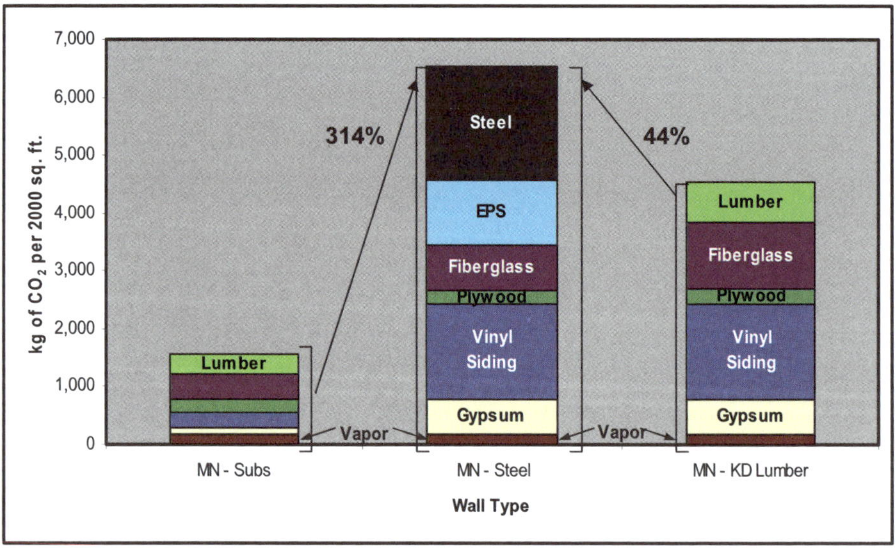

Figure 3-4. Global warming potential of alternative wall-framing materials, Minneapolis (Source: Lippke and Edmonds 2006, 61).

*This corrects the reference to "lower red line" in Malmsheimer et al. (2008) at page 134.

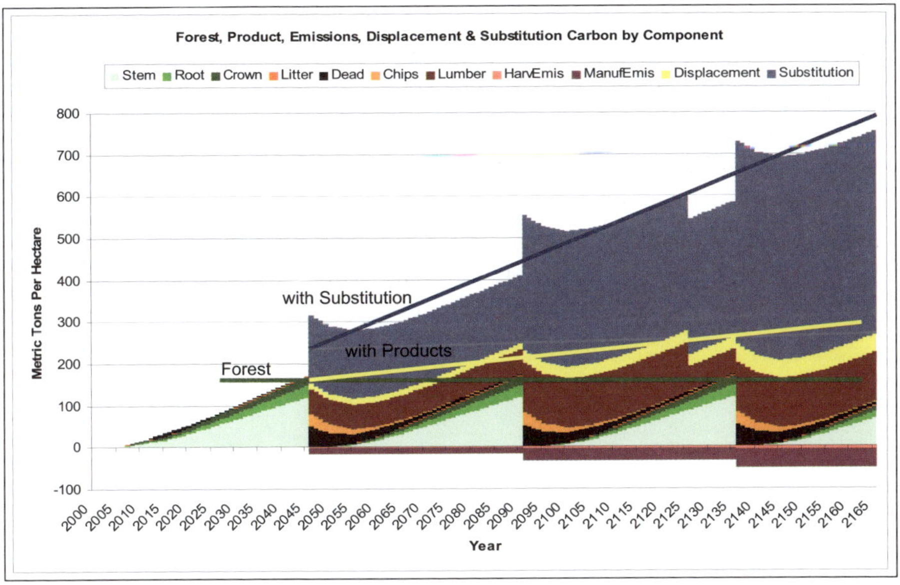

Figure 3-5. Carbon in forest, product, fuel displacement, and fossil fuel–intensive product substitution pools (Source: Perez-Garia et al. 2005).

fuel–intensive building materials like concrete and steel framing (black* line), emissions are avoided. The combined pools of carbon stored in the forest, forest products (net of processing, including the bioenergy from bark, or hog fuel, from mill waste), and avoided fossil fuel–intensive substitutes increase over time—with important consequences for carbon policy (USFS 2005).

CORRIM has also conducted life-cycle assessments for different kinds of wood products. Plywood sheathing has a 3 percent lower environmental impact in a completed house than oriented-strand board (OSB) (although OSB has fewer water-related environmental consequences, probably because at the time of the research, some OSB mills were in compliance with new, stricter water quality standards) (Lippke et al. 2004). Conversely, substituting wood dimension joists for engineered I-joists results in little difference in the environmental performance indices because the greater material efficiency of the I-joists is offset by the increased use of resins and energy (Lippke et al. 2004). However, material use efficiency is by itself very important, since only half as much fiber is used for engineered I-joists as for the equivalent dimension lumber joints.

*This corrects the reference to "upper red line" in Malmsheimer et al. (2008) at page 134.

Forest Rotations and Conversion

The sooner wood products can be produced from forests, the sooner they can displace the emissions from fossil fuel–intensive products. Thus, intensive, short-term commercial rotations, while storing less overall carbon in the forest, result in lower carbon emissions when life-cycle assessments include forest and product carbon storage as well as the emissions from substitute products. Some estimates indicate that a forest managed for wood production will provide a net sequestration at least double that of an unmanaged forest in the Pacific Northwest (B. Lippke, University of Washington, pers. comm., August 2007). If, however, the goal is to sequester carbon in the forest, management for long rotation and old-growth will lead to significant ecosystem carbon storage (Harmon et al. 1990).

Carbon stocks are affected by changes in land use. When forestland is converted to nonforest use, the carbon stored both on that land and in its wood products is lost, along with those products' potential to be substituted for fossil fuel–intensive products.

Unnaturally high fuel loads in many forests provide wood substitution opportunities. Thinning heavily stocked stands and using the wood in long-lived products or converting it into biofuel would avoid the carbon emissions associated with fossil fuels and fossil fuel–intensive products. These same areas would thus contribute to reduced GHG emissions and increased carbon storage in wood products, without the risks of GHG releases that overstocking can create if the forest burns (Chapter 5 addresses how GHG emissions can be prevented through wildfire behavior modification). Simulations have shown the reduction in emissions from carbon stored in products and displacement of fossil fuel–intensive products and fuels can be as much as five times larger than the carbon stored in Inland West forests in 100 years (Oneil 2007).

Wood Substitution Climate Change Policy

Policies intended to slow global warming can easily have unintended consequences. A policy that lowers the cost of wood, for example, would motivate builders and consumers to select wood framing and floors in residential construction. As the demand for wood rises, relative to fossil fuel–intensive materials, more investment in growing wood for this market would occur, resulting in further reductions in emissions. However, if a carbon credit is given only to growing trees in forests, it would likely lengthen rotations, reduce the production of wood products, and possibly increase the use of fossil fuel–intensive products, thereby increasing GHG emissions (B. Lippke, University of Washington, pers. comm., November 2007). Developing carbon credit markets that motivate true reductions in carbon emissions must address all carbon pools and their GHG emissions. Such markets will not be successful if they focus only on carbon stored in forests or a single stage of processing.

Measuring the life-cycle inventory of environmental impacts and assessing their effects across all stages of processing are critical to evaluating the consequences of different processes, product uses and designs, and forest management. The values (costs) of these impacts must be accurately reflected in the market if we want to motivate the changes in consumption and investments that will reduce carbon emissions. As an example, the Swedish parliament has recognized an opportunity to reduce GHG emissions by reducing the use of concrete in buildings and has instituted policies, educational campaigns, regulations, and building codes to promote the use of wood (Sathre 2007).

Although wood product substitution does not permanently eliminate carbon from the atmosphere, it does sequester carbon for long durations and can offset the use of more GHG-intensive products. When wood is harvested and used to make lumber, furniture, plywood, or other wood products, carbon is sequestered for the life of the given wood product. Once the wood product has served its useful life, landfill management techniques can further delay the conversion of wood to GHG emissions, or the wood can be used for power generation (offsetting generation by fossil fuel–fired power plants) or recycled into other potentially long-lived wood products. Regardless of the particular pathway followed after a product's useful life, wood substitution is a viable and important technique to immediately address climate by preventing GHG emissions.

Chapter 4

Preventing GHG Emissions through Biomass Substitution

GHG emissions can be reduced through the substitution of biomass for fossil fuels that emit more GHG per functional unit. The production and use of biomass fuels and bio-based products is one way to reduce oil and gas imports and improve environmental quality. Biomass can be used as an offset for fossil fuels like coal, natural gas, gasoline, diesel oil, and fuel oil. At the same time, such uses can enhance domestic economic development by supporting rural economies and fostering new industries making a variety of renewable fuels, chemicals, and other bio-based products (California Biomass Collaborative 2005; English et al. 2006; J. R. Smith et al. 2007).

Biomass is the largest domestic source of renewable energy, providing 3.227 quadrillion BTUs (quads) or approximately 48 percent of the nation's renewable energy (EIA 2006). Of the 3.227 quads of biomass energy used in 2005, 2.114 quads (65 percent) came from wood. Of a total of 1.875 quads of industrial biomass energy in 2005, 1.460 quads, almost 88 percent, was used by forest industries, such as sawmills, oriented strand board mills, and pulp and paper mills (EIA 2006). Most of the renewable energy used by forest industries comes either from their own industrial plant residuals or from wood residues purchased from other wood using industries. Zerbe (2006) estimates that up to 10 percent of our nation's energy requirement could eventually be produced from wood, compared with the 3 percent we currently produce.

Studies of conversion technologies show that 1 dry ton of forest waste can be converted to 75 to 85 gallons of ethanol fuel or 550 to 650 kilowatt-hours of electricity. If only 30 percent of the estimated 368 million dry tons of forest waste available in the United States each year were in a suitable form and concentration to be converted to energy, these wastes could produce 9.2 billion to 10.4 billion gallons of ethanol or 67 billion to 80 billion kilowatt-hours of electricity (Perlack et al. 2005) (See Table 4-1).

Table 4-1. Biomass conversion factors.

1 green ton of chips	= 2,000 lbs. (not adjusted for moisture)
1 Bone Dry Ton (BDT) of chips	= 2 green tons (assuming 50% moisture content)
10,000 lbs. of steam	= 1 megawatt hour (MWH) of electricity
1 Megawatt (MW)	= 1,000 horsepower
1 MW	= power for approximately 750-1,000 homes

Note: A 50 MW biomass powerplant will use 1,200 BDT/day; 100 chip vans/day
Source: Adapted from TSS Consultants 2006.

Bioenergy Basics

Technologies for Converting Wood to Energy. The technologies for converting woody biomass to energy include direct burning, hydrolysis and fermentation, pyrolysis, gasification, charcoal, and pellets and briquettes (Bergman and Zerbe 2004; Zerbe 1983, 2006).

Direct burning. The most effective way to use woody biomass for energy is to burn it in a combustion system, such as a boiler, fitted with emissions controls. Net boiler efficiencies range from 60 percent for greenwood at 60 percent moisture content to 80 percent for oven-dried wood. Wood can also be cofired with coal or natural gas.

Hydrolysis and fermentation. In the pulp and paper industry, the hemicellulosic materials from wood can be extracted at the beginning of the process via hydrolysis and then fermented using enzymes to produce ethanol and other products. The remaining cellulosic materials are still available for producing pulp and paper.

Pyrolysis. The heating of wood with limited or no oxygen to prevent combustion, called pyrolysis, produces liquid fuel, char, and gas. Lower temperatures produce higher portions of liquid and char; higher temperatures produce more gas. Another process, "flash pyrolysis," produces liquid bio-oil. In flash pyrolysis, biomass is heated rapidly to 400° to 600°C in the absence of air, with 70 to 75 percent of the feedstock converted into bio-oil. The oil is somewhat corrosive, but it can be used as boiler fuel or, with subsequent treatment, diesel fuel.

Gasification. Gasification uses oxygen and heat to produce a synthetic gas ("syngas") from biomass. This process was used during World War II and earlier, when crude oil supplies were limited. Gasification can be used to power internal combustion engines or gas turbines to drive electrical generators. Energy efficiencies from gasification for generating electricity range from 22 to 37 percent, compared with 15 to 18 percent for steam produced from combustion.

Charcoal. The production of charcoal is a pyrolytic process. Charcoal is made by heating wood in airtight ovens or retorts, or in kilns supplied with limited amounts of air. The heat breaks down the wood into gases, a tar mixture (lignosulfonic acid), and charcoal. The potential fuel yield is only about half of the original energy content of the wood.

Pellets and briquettes. Wood pellets and briquettes are more fully processed and refined than chips, sawdust, chunkwood, and other forms of solid wood and are more uniform in size and physical properties, such as ash content. Wood pellets are easily combusted using sophisticated stoves or burners with automatically controlled feeder systems. Premium wood pellets burn at an efficiency of 83 percent, which offsets the extra energy used in making them (Bergman 2004).

Energy Uses for Wood. Energy uses for wood include thermal energy for steam, heating, and cooling; electrical generation and cogeneration; and transportation fuels.

Thermal energy. Installations for converting wood into thermal energy for space heating and cooling generally involve four size units.

- *Micro scale:* Up to 1 megawatt (MW) for residences or schools. This can involve firewood furnaces or gasification units and the use of a boiler to produce warm air or hot water for pipe heating systems.
- *Small scale:* 1 to 5 MW to produce high-pressure steam for heating or to energize an air-conditioning system.
- *Medium scale:* 5 to 15 MW for larger institutions, such as community colleges or hospitals, involving various types of combustors and boilers.
- *Large scale:* More than 15 MW. These systems are common in the forest products industry, most commonly in dry kilns.

Electrical generation and cogeneration. Wood can be used to generate electrical power from steam-driven turbine generators or gas turbines. Most wood-powered plants in the United States are in the 10 to 20MW range, but some are larger than 70MW. The average biomass-to-electricity efficiency of the industry is 20 percent. The nearest-term low-cost option for using biomass in power generation is cofiring with coal (Bain and Overend 2002). Cogeneration or combined heat and power are a more efficient use of wood than for the production of electricity alone. Wood also has the potential to be used for fuel cells; the wood is converted to hydrogen, methanol, or ethanol to power fuel cells.

Transportation fuels. Ethanol and other transportation fuels can be produced from almost any source of woody biomass. Methanol, another liquid fuel, can be made from wood as an alternative to gasoline or diesel. Even gasoline can be made from wood, but this requires gasification of wood and its conversion to syngas. The most direct way of making gasoline and diesel from woody biomass or other organic feedstocks is through what is known as the Fischer-Tropsch (FT) gasification process.

Total Forest Biorefinery Concept. Currently, the Department of Energy and the forest products industry are looking at the potential for pulp mills to become a "total forest biorefinery" (Larson et al. 2006) (Figure 4-1). Some of the hemicelluloses from wood chips would be extracted prior to pulping and converted (by hydrolysis) into wood sugars, which can be fermented into ethanol and produce xylitol and acetic acid. The process would divert hemicelluloses and acetic acid from direct combustion into valuable byproducts without significantly reducing the yield of cellulose pulp.

Preliminary studies indicate that the process may be economically feasible and could add to the output of ethanol as a transportation fuel while enhancing economic returns for pulp mills. Another component of the total forest biorefinery concept is the gasification of spent pulping "black liquor,"

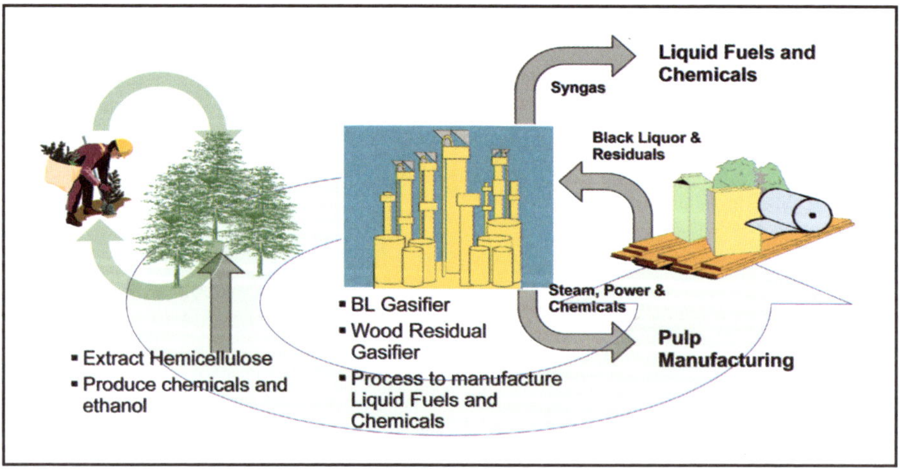

Figure 4-1. Total biorefinery concept applied to pulp and paper industry (Source: Pacheco 2005).

which is conventionally burned via direct combustion in a Tomlinson recovery boiler in pulp mills. Gasification would be used to produce a syngas (H_2 and CO) that could then be reformed by a catalytic process into various chemicals and transportation fuels (ethanol, methanol, dimethyl ether, and FT diesel).

The latter technology is still being refined and perfected. A recent report showed the potential to displace 2.2 billion barrels of oil annually, with an additional benefit of cutting approximately 91 million tonnes of carbon emissions annually, if the total forest biorefinery concept were adopted by the nation's kraft pulp and paper industry. A fully developed pulp mill biorefinery industry could double or more the liquid fuel production of the current corn-based ethanol industry in the United States (Larson et al. 2006).

Biomass Energy Production and Greenhouse Gas Emissions

The use of wood to produce energy opens two opportunities to reduce GHG emissions. One involves using forest biomass for electrical power generation, rather than allowing it to accumulate and decay on site or removing it by open field burning. The other is the substitution of woody biomass as an energy source in place of fossil fuels.

Wood Burning and Greenhouse Gas Emissions. Biomass for power generation results in a 98.4 percent reduction in emissions compared with open field burning (Table 4-2) (Darley 1979). These ranged from an 84.8 percent reduction for nitrogen oxides to a 100 percent elimination of hydrocarbons.

Hasse (2007) compared emissions from biomass boilers with emissions from pile burning, prescribed burning, and forest fires (Table 4-3). He found

Table 4-2. Open field burning versus biomass boiler emissions.

Pollutant	Open-field burning (lbs./ton)	Biomass boiler (lbs./ton)	Percentage reduction for biomass boiler
Sulfur oxides	1.7	0.04	97.6
Nitrogen oxides	4.6	0.70	84.8
Carbon monoxide	70.3	0.40	99.4
Particulates	4.4	0.26	94.1
Hydrocarbons	6.3	0.00	100.0
Total	87.3	1.40	98.4

Source: Darley 1979.

Table 4-3. Pile burning, prescribed burning, and forest fire versus biomass boiler emissions.

	Pounds of emissions per green ton			
Disposal method	**PM10**	**NO_x**	**VOC**	**CO**
Pile burning[a]	19 to 30	3.5	8 to 21	54 to 312
Prescribed burning[b]	24	4.0	13	224
Forest fire[b]	15	4.0	21	140
Biomass boiler[c]	2.1	2.8	0.6	1.7
Average reduction (%)	89	30	96	99

[a] Werner (2000), available at http://www.arb.ca.gov/ei/see/memo_ag_emission_factors.pdf.
[b] Environment Australia, Emissions Estimation Technique Manual for Aggregated Emissions from Prescribed Burning and Wildfires,
Version 1.0, September 1999.
[c] Based on Chiptec gasifier; other systems are similar.
Source: Hasse 2007.

a 99 percent reduction in carbon monoxide emissions, 30 percent for nitrous oxides, 96 percent for volatile organic compounds, and 89 percent reduction for PM10 particulates. Emissions from open burning also include methane (CH_4), which has a global warming potential of 23 (i.e., 1 pound of CH_4 emissions is equivalent to 23 pounds of CO_2).

It is estimated that the United States needs to build 1,200 new 300-megawatt power plants during the next 25 years just to keep pace with projected increases in demand for electricity (Hasse 2007). Coal, the most abundant energy source available in the United States, will likely continue to be a major source of energy for electricity production. Electricity-generating plants are already the largest stationary source of GHG emissions from fossil fuels. How can the nation meet its energy needs without exacerbating air pollution and GHG emissions?

Although some energy needs can be met by renewable sources such as solar and wind, biomass must play a crucial role. Woody biomass, used as a feedstock to be burned or mixed with coal, presents a viable short- and mid-term solution to low-cost and large-scale alternative energy feedstocks. Cofiring wood with high-sulfur coal reduces sulfur air emissions and problems with mercury and other heavy metals. Cofiring woody biomass with coal on a 5 to 10 percent energy basis and using biomass with coal to produce liquid fuels are two possible clean energy solutions. Cofiring woody biomass with coal could provide a major increase in the demand for woody biomass for energy production. Woody biomass can also be added to oil-

and gas-generated electric production processes to reduce GHG emissions (Morris 2007).

Wood-Based Liquid Fuels and Greenhouse Gas Emissions. Annual US gasoline consumption today is 140 billion gallons, and US diesel fuel consumption is 56 billion gallons. Each year the nation uses 6.5 billion barrels of oil but produces only 2.5 billion barrels of oil from domestic sources. That means that 4.0 billion barrels of oil has to come from foreign sources—and often from volatile parts of the world—to meet annual needs (Hasse 2007).

In February 2006, President Bush announced the Advanced Energy Initiative, designed to make cellulosic ethanol cost competitive with corn by 2012. The initiative has two goals:

- *"20 in 10"*: replace 20 percent of today's gasoline usage in 2010 with biofuels.
- *"30 in 30"*: replace 30 percent of today's gasoline usage in 2030 with biofuels.

In his 2007 State of the Union address, President Bush called for a mandatory 35 billion-gallon renewable fuel standard by 2017. A June 2007 Government Accountability Office report calculated 2006 ethanol and biodiesel production at 4.9 billion gallons a year, or 3 percent of the current US demand. It also estimated that the maximum annual production from corn ethanol would be 15 billion to 16 billion gallons, and from biodiesel, 2 billion gallons (Hasse 2007). This leaves an annual gap of 17 billion to 18 billion gallons of transportation fuels that will have to come from cellulosic and other feedstocks to meet the 35-billiongallon renewable fuel standard. The gap could be filled by cellulosic ethanol made from wood. Given that 1 dry ton of forest waste can be converted to 75 to 85 gallons of ethanol fuel, 30 percent of the estimated 368 million dry tons of available forest residues could produce 9.2 billion to 10.4 billion gallons of ethanol (Perlack et al. 2005). If 60 percent of the residues were available to make cellulosic ethanol, potential production would be in the range of 18 billion to 20 billion gallons, making the President's mandatory renewable fuel standard of 35 billion gallons of renewable fuels achievable.

At present, increasing amounts of federal funding and venture capital are being channeled into the production of cellulosic ethanol. This is being driven by national security concerns about the increasing US reliance on foreign crude oil, concerns over greenhouse gas emissions and global warming, the realization that corn-based ethanol production will likely peak at 15 billion to 20 billion gallons by 2030, and associated economic development opportunities.

One major challenge in making cellulosic-based fuels is the development of improved technologies to reduce production costs. Another involves supply and demand: the production of renewable transportation fuels from cellulosic feedstocks could affect domestic supplies and costs for existing feed and fiber uses.

The production of cellulosic ethanol has been a subject of studies related to energy conversion efficiency. For the most part, studies show positive energy input-output ratios ranging from 4.40 to 6.61 (Tyson et al. 1993; Lynd and Wang 2004; Sheehan et al. 2004). The only exception has been a study by Pimentel and Patzek (2005), who report a negative energy ratio of 0.69. The difference stems from the assumption by Pimentel and Patzek that industrial process energy is generated by fossil fuel combustion and electricity rather than lignin combustion (Hammerschlag 2006). In most models, cellulosic production generates industrial energy with lignin combustion rather than fossil fuels and electricity, and thus fossil energy inputs are consistently far less than the energy value of ethanol and surplus electricity delivered. Hammerschlag (2006) also notes that cellulosic fuel is a developing industry, and more mature processes with considerably greater ratios of energy outputs to inputs are possible.

The National Renewable Energy Laboratory has compared the energy required to produce gasoline, corn ethanol, and cellulosic ethanol, based on data by Wang et al. (2005) (Figure 4-2), and Roj (2005) has compared the energy efficiencies and greenhouse gas emissions from fossil and renewable fuels (Figure 4-3).

In 2007, the Environmental Protection Agency's Office of Transportation and Air Quality estimated the percentage change in life-cycle GHG emissions, relative to the petroleum fuel that is displaced, by a range of alternative and renewable fuels (Figure 4-4). The fuels are compared on an energy equivalent or BTU basis. For instance, for every BTU of gasoline that is replaced by cellulosic ethanol, total life-cycle GHG emissions would be reduced by 90.9 percent. These emissions account not only for CO_2 but also for methane and nitrous oxide. The cellulosic ethanol estimate represents an average mix of feedstock sources (including hybrid poplar, switchgrass, and corn stover) to produce ethanol through two production processes (a fermentation process, and ethanol produced from forest waste via gasification).

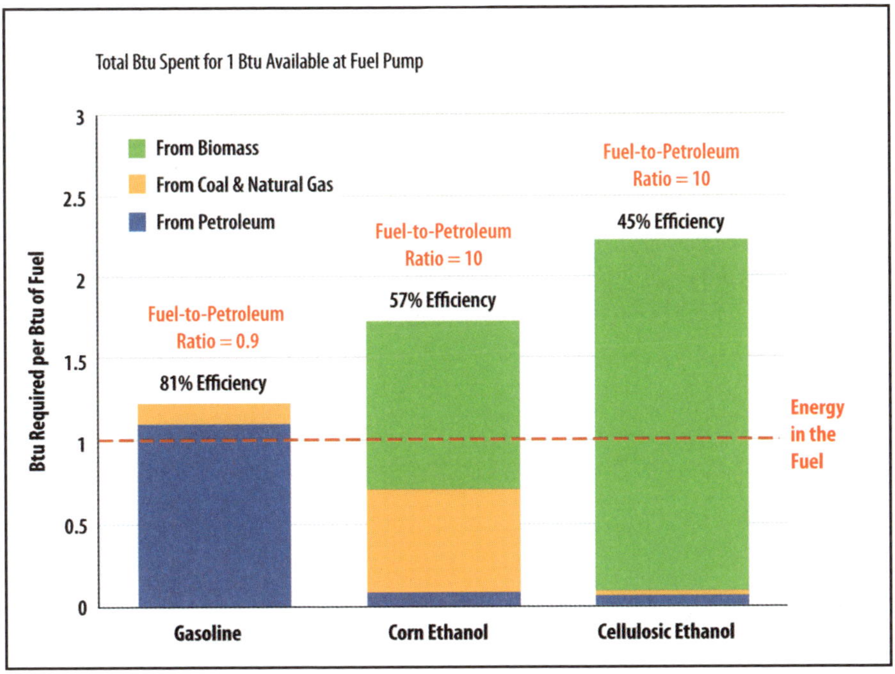

Figure 4-2. Energy required to produce fuels (Source: NREL 2006).

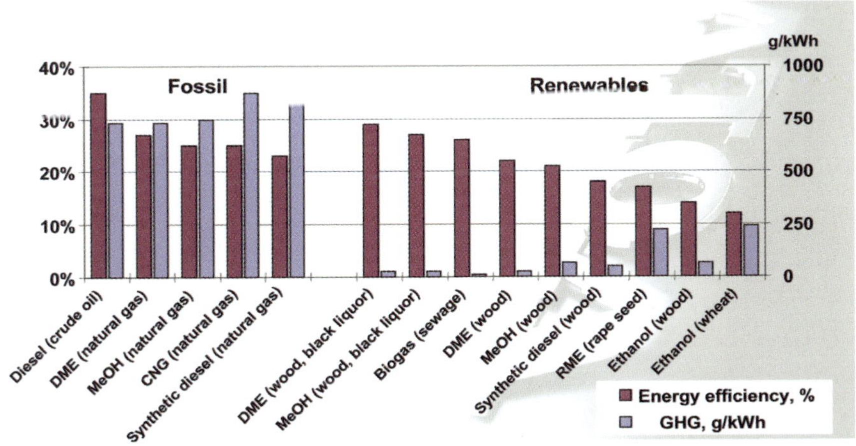

Figure 4-3. "Well to Wheels" analysis of energy efficiency and greenhouse gas emissions (Source: Roj 2005).

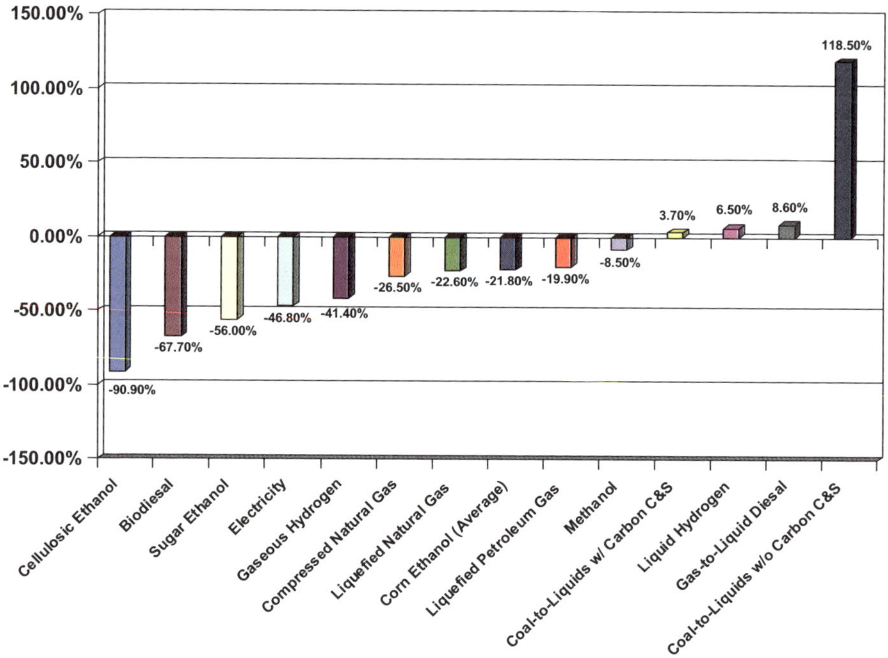

Figure 4-4. Life-cycle greenhouse gas emissions for renewable fuels compared with traditional gasoline (Source: US EPA 2007a).

Woody Biomass Feedstocks and Their Availability

The research illustrated in Tables 4-1 and 4-2 and Figures 4-1, 4-2, and 4-3 demonstrate that substituting cellulosic biomass for fossil fuels greatly reduces GHG emissions. But is sufficient woody biomass available to address US energy needs?

Woody biomass essentially is any tree or part thereof and any associated woody plant materials. It includes wood from the bole (trunk) of the tree, limbs, tops, roots, and even the foliage. It includes trees that have been killed or damaged by fire, insects, diseases, drought, or wind or ice storms. It can also include trees that have been grown specifically for the production of energy wood—dedicated short-rotation tree or woody crops—and trees removed for fuel reduction, restoration, or other cultural treatments. In its broadest sense, woody biomass also includes raw materials as well as postconsumer recycled paper and wood products.

Nonmerchantable forest wastes and low-value trees can serve as a source for bioenergy feedstock, but there are infrastructure and sustainability challenges associated with the collection of these feedstocks. Collection and transportation costs of woody biomass can be significant and vary greatly from region to region. Although larger trees are generally more cost-effective to harvest and use, such trees usually have a higher value for traditional forest products, such as sawtimber, pulpwood, and manufactured panels.

Perlack et al. (2005) estimated that the United States can produce 1.3 billion dry tons of biomass annually on a sustainable basis. The Department of Energy's National Renewable Energy Laboratory and the US Forest Service estimate that 1.3 billion dry tons would roughly yield an energy heating value of 3.5 billion barrels of oil—equivalent to the US domestic oil production in 1970, the peak year of domestic oil production (Figure 4-5). The woody biomass component of these 1.3 billion dry tons is estimated to be 368 million dry tons. Perlack et al. (2005) note that this annual sustainable biomass estimate is conservative. The calculations exclude all protected wilderness and roadless areas, steep slopes, environmentally sensitive areas, and areas where regeneration would be difficult. Wood considered merchantable for other products was not counted, and the figure also accounts for physical limitations of on-site recovery and leaving sufficient woody debris on site to alleviate potential adverse effects on soil and water quality.

The estimated 368 million dry tons of annual sustainable woody biomass available in the United States comes from several sources:

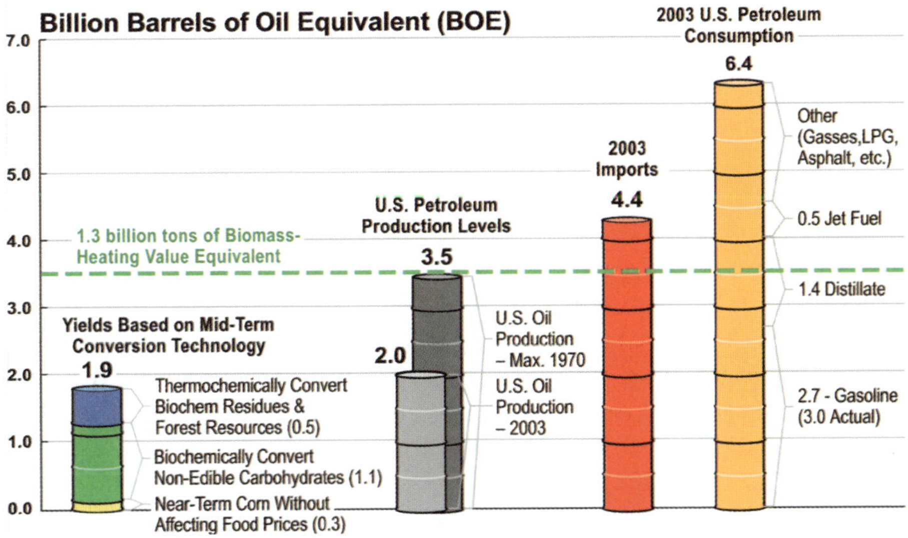

Figure 4-5. Heating value equivalent of biomass compared with oil production and consumption (Source: Pacheco 2005).

- logging and other residues (41 million dry tons);
- fuel treatments (60 million dry tons);
- fuelwood (35 million dry tons);
- forest products industry wastes (106 million dry tons);
- urban wood residues (37 million dry tons); and
- forest growth (89 million dry tons).

The last category, forest growth, warrants further discussion. The Fifth Resources Planning Act Timber Assessment of the US Forest Service projects the continued expansion of standing forest inventory despite an estimated conversion of about 23 million acres of forestland to other uses (Haynes 2003). The size of the standing forest inventory will increase because annual forest growth will continue to exceed annual harvests and other removals from the inventory. At the same time, the forest products industry will continue to become more efficient in the way it harvests and processes wood products. The demand for forest products is also projected to increase more slowly than in the past because of a general declining trend in the use of paper and paperboard products relative to GNP and the relatively stable forecast of housing starts (Perlack et al. 2005).

The Department of Energy and USDA analyses did not include wood that is currently merchantable at the lower size and quality specifications for conventional products, such as pulpwood and small sawlogs. Depending on local market conditions (e.g., low-price wood and/or high-price oil markets), this wood could be an additional resource for bioenergy and bio-based products. For example, the US South has vast forests that are being commercially thinned to improve stand quality. It is projected that approximately 8 million dry tons could be available annually from these treatments (Perlack et al. 2005). The reduction in pulp utilization in the United States resulting from the globalization of pulp production may make even more such thinnings available for energy in the future.

One forest management option for increasing the production of woody biomass is short-rotation energy crops using rapid growing species such as alder, cottonwood, hybrid poplar, sweetgum, sycamore, willow, and pine. Perlack et al. (2005) did not count short-rotation tree energy crop production potential or account for possible production increases achievable through genetics or more intensive silvicultural practices. A yield figure of 8 dry tons per acre would add approximately 10 million dry tons annually to the estimated 368 million* dry tons of US woody biomass production.

*This corrects the "368 billion" figure in Malmsheimer et al. (2008) at page 140.

Chapter 5

Preventing GHG Emissions through Wildfire Behavior Modification

Wildland fires are a major contributor to national and international greenhouse gas emissions, adding as much as 126.4 million tonnes of carbon dioxide emissions in the United States during 2005 (US EPA 2007b). Active forest and wildland fire management strategies can dramatically reduce CO_2 emissions while also conserving wildlife habitat, preserving recreational, scenic, and wood product values, and reducing the threat of wildfires to communities and critical infrastructure.

Wildfire GHG Emissions

Smoke from wildfires emits particulates, CO_2, and other GHGs such as methane. The Environmental Protection Agency estimates that forest wildfire emissions in the lower 48 states and Alaska released an average of 105.5 million tonnes (range, 65.3 to 152.8) of carbon dioxide into the air each year from 2000 to 2005 (US EPA 2007b). Another study indicates that annual wildfire CO_2 emissions from 2002 to 2006 may actually average as high as 293 million tonnes per year, a major portion of which comes from forests (Wiedinmyer and Neff 2007).

To take one example, the July 2007 Angora wildfire in South Lake Tahoe affected only 3,100 acres of forestland, yet it released an estimated 141,000 tonnes of carbon dioxide and other GHGs into the atmosphere, and the decay of the trees killed by the fire could bring total emissions to 518,000 tonnes (Bonnicksen 2008). This is equivalent to the GHG emissions generated annually by 105,500 cars. In another example, Bonnicksen (2008) found that four California wildfires emitted an average 65 tonnes of greenhouse gases per acre and that with the release of CO_2 from decay over the next 100 years, the 144,825 burned acres will emit 35 million tonnes of greenhouse gases—equivalent to the annual emissions from half of California's 14 million cars. In 2006, wildfires burned nearly 10 million acres in the United States.

The Intergovernmental Panel on Climate Change describes the global impacts of smoke:

> Destruction of forest biomass by burning releases large quantities of CO_2 and is estimated to create 10 percent of annual global methane emissions as well as 10–20 percent of global NO_2 emissions. Thus, fire can have a significant effect on atmospheric chemistry (IPCC, 1992). The process is well known in terms of general effects, but it has many uncertain parameters in relation to specific fire events because fire effects are related to fuel amounts, arrangements, and conditions as well as weather conditions at the time of combustion—all of which can be highly variable or unpredictable (Goldammer, 1990; Dixon and Krankina, 1993; Price et al., 1998; Neuenschwander et al., 2000). (Sampson and Scholes 2000, 271)

The effect of particulates on climate change is uncertain (Kaufman et al. 2005). Some scientists contend that smoke reflects sunlight and reduces surface temperatures (Pearce 2005); others consider this phenomenon only temporary or transitory and say that long-term warming can result (Cess et al. 1985); still others believe that smoke may provide cooling in lower latitudes but warming in higher latitudes (R. Neilson, US Forest Service, pers. comm., October 2, 2007).

Wildfires in the United States and in many other parts of the world have been increasing in size and severity, and thus future wildfire emissions are likely to exceed current levels. Three strategies to reduce wildfires and their GHG emissions can address this trend:

- pretreatment of fuel reduction areas—that is, removing some biomass before using prescribed fire;
- smoke management—that is, adjusting the seasonal and daily timing of burns and using relative low-severity prescribed fires to reduce fuel consumption; and
- harvesting small woody biomass for energy, or removing larger woody material (over 4 inches in diameter) for traditional forest products and burning residuals.

Removing materials for bioenergy applications (described in Chapter 4) can reduce the threat of catastrophic wildfires and the net smoke and GHG emissions. In addition, active management of forest landscapes has the potential to decrease the area burned in catastrophic wildfires by 50 to 60 percent (Finney 2000). This reduces soil erosion and related watershed problems.

Prescribed fire managers follow stringent air quality and burn plan requirements. In addition to detailed weather and fuel modeling, prescribed burn emissions must comply with federal and state air quality requirements. These requirements include maximum allowable concentrations of the nine pollutants regulated by National Ambient Air Quality Standards (The Nature Conservancy n.d.; see also US EPA 2007c).

To qualify for emission reduction credits for prescribed burns, managers must comply with federal and state emission and smoke reduction standards. The Clean Air Act requires that emission reductions be real, quantifiable, permanent, verifiable, and enforceable. Some states (e.g., California, Florida, and Montana) have developed their own guidelines; however, the Environmental Protection Agency has published only interim federal rules.

Wildfire and Climate Trends

Catastrophic wildland fires in the United States during the past decade have added tens of millions of tonnes of carbon dioxide and greenhouse gases to the atmosphere each year. Climate change, rural housing development, and human encroachment into wildlands will only exacerbate the problem (Field et al. 2007). Wildfires in the new millennium have even prompted new terms to categorize wildfires that are far beyond the scale of conflagrations in recent human history. "Megafire," for example, refers to one very large fire or a group of fires that burn into a single fire; "fire complex" refers to a series of fires in a short period of time within a specific area that are managed as one large fire.

Since 2000, at least 12 states have experienced the largest wildfires in their modern history (Table 5-1). Six of the worst fire seasons (including 2007) in the past 47 years, based on area burned, have occurred since 2000. Reduced rainfall and changes in seasonal weather patterns—primarily warmer, drier air masses—have influenced wildfire behavior. For example, the 2006 fire season started in January—an unusual time of year for catastrophic wildfires—when more than a million acres burned in Texas and Oklahoma, and extended drought and hot, dry weather in Georgia and Florida caused record fires from mid-April until July 2007.

Figure 5-1 shows the effectiveness of prevention, presuppression, and other efforts in reducing the number of fires since the mid-1980s. However, it also illustrates how increased fuel loads, climate change, and other factors have increased the total area burned, which indicates an increase in megafires. Drought and climate change may increase the risk of insect and disease

Table 5-1. Largest fires in state history since 1960

Year	Fire	Location	Size (acres)
2004	Taylor Complex	Alaska	1,305,592
2006	East Amarillo Complex	Texas	907,245
2005	Southern Nevada Complex	Nevada	508,751
2002	Biscuit	Oregon	499,570
2002	Rodeo-Chediski	Arizona	468,638
2007	Murphy Complex[a]	Idaho	464,702
2007	Georgia Bay Complex[b]	Georgia	441,705
2007	Milford Flat	Utah	363,052
2000	Valley Complex (Bitterroot)	Montana	292,070
2003	Cedar	California	279,246
2000	24 Command	Washington	162,500
2002	Hayman	Colorado	137,760
2000	Kate's Basin	Wyoming	137,600

[a] The Murphy Complex burned a total of 653,100 acres in Idaho and Nevada.
[b] The Georgia Bay Complex burned a total of 564,450 acres in Georgia and Florida.
Source: Compiled from National Interagency Fire Center 2007.

epidemics, killing or weakening trees and adding to the dead fuel component. This often increases fire intensity and the GHG emissions released. Large-scale insect infestations can affect fire suppression tactics and firefighter safety because fuel loads have changed, increasing spotting potential and altering fire behavior and fireline intensity. Climate and weather influences will further complicate suppression challenges.

Virtually all climate change models forecast an increase in wildfire activity, although IPCC cautions that "fire, insects and extreme events are not well modeled" (Easterling et al. 2007, 290). IPCC notes an increase in North American wildfires attributable, with "high confidence," to climate change: "the forested area burned in the western U.S. from 1987 to 2003 is 6.7 times the area burned from 1970 to 1986" (Field et al. 2007, 623, citing Westerling et al. 2006).

Even with a stable climate, the area burned and threats to humans may continue to increase with fuel buildup and human presence in wildlands. Encroachment and development, the proximity of population centers to wildlands, and more human caused fires (both arson and accidental) all significantly increase the risk and consequences of wildfire, including the release of GHGs. It will take many years to reduce the tremendous fuel buildup in dry forest systems (such as ponderosa pine) whose historic fire regimes, characterized by frequent low-intensity fires, were interrupted by more than a century of wildfire suppression, grazing, logging, and a cooler and moister climate in the middle 1900s. Community wildfire protection planning, as authorized in the Healthy Forests Restoration Act of 2003 (P.L. 108–148) and described by the Society of American Foresters (2004), can address this problem. However, under extreme fire behavior scenarios, which could be exacerbated by climate change, increased accumulations of hazardous fuels will cause ever-larger wildfires.

Not all climate models paint a bleak future for forests. Research by the US Forest Service's Pacific Northwest Research Station indicates that the increase in precipitation associated with climate change may moderate fire behavior, even as the fire season lengthens and temperatures rise. Using the Mapped Atmosphere-Plant-Soil System (MAPSS) computer model, the Forest Service forecasts an increase in western woody and grass fuels (carbon capture) for the 21st century (USFS 2004). The pinion-juniper forest type has already expanded its range in the West. MAPSS models predict a dramatic shift to coniferous forests (mostly dry forest types, such as ponderosa pine) over the

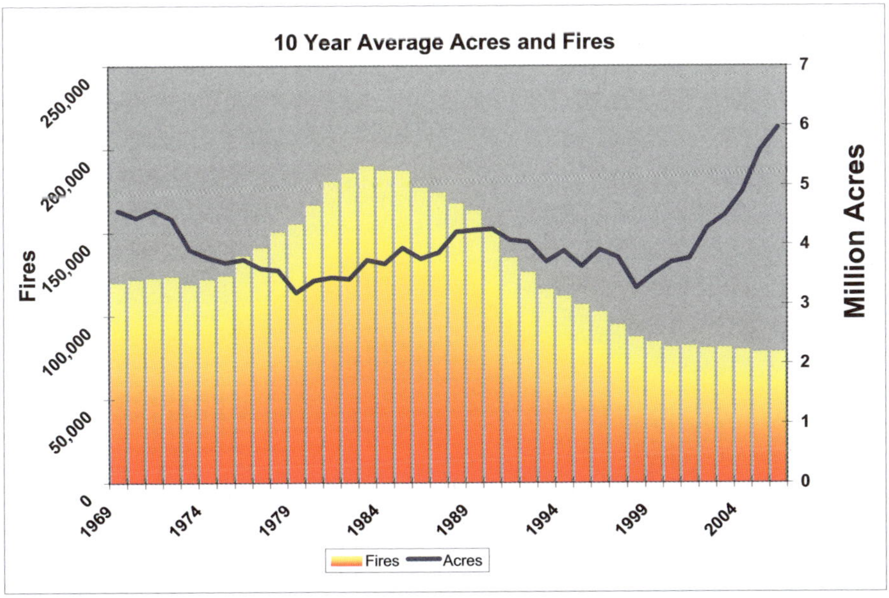

Figure 5-1. Ten-year averages of acres burned and number of fires (Source: Compiled from National Interagency Fire Center 2007).

coming century. The forecasted changes for the East, however, under the MAPSS "moderate" climate scenario—a rise in surface temperature of 4.2°C (7.5°F) and an 18 percent increase in precipitation—reach a "tipping point" in carbon balance. Under the MAPSS "considerable warming" scenario—5°C (9°F) and a 22 percent increase in precipitation—a dramatic decrease in vegetation density is predicted to result in a net carbon loss for the United States. However, climate projections differ; for example, the most recent IPCC scenarios project increasing drought in the US Southwest (IPCC 2007).

Wildfire is not unique to North America, of course. For example, in summer 2007, international attention was focused on Greece when wildfire killed 64 people, burned more than 450,000 acres, and shrouded ancient ruins, such as the site of the original Olympics, in smoke. Though they received more media attention, these fires added much less CO_2 to the atmosphere than the estimated 28.9 million acres (11.7 million hectares) of wildfires that burned in the Russian Federation that same year. Figure 5-2 illustrates how climate change may increase the amount of biomass burned by wildfires (Neilson 2007). A 2002 international assessment estimated the 1998 wildfires in Siberia "released close to 180 million ton[ne]s of carbon to the atmosphere which contributed to the formation of 520 million t[onnes] of carbon dioxide,

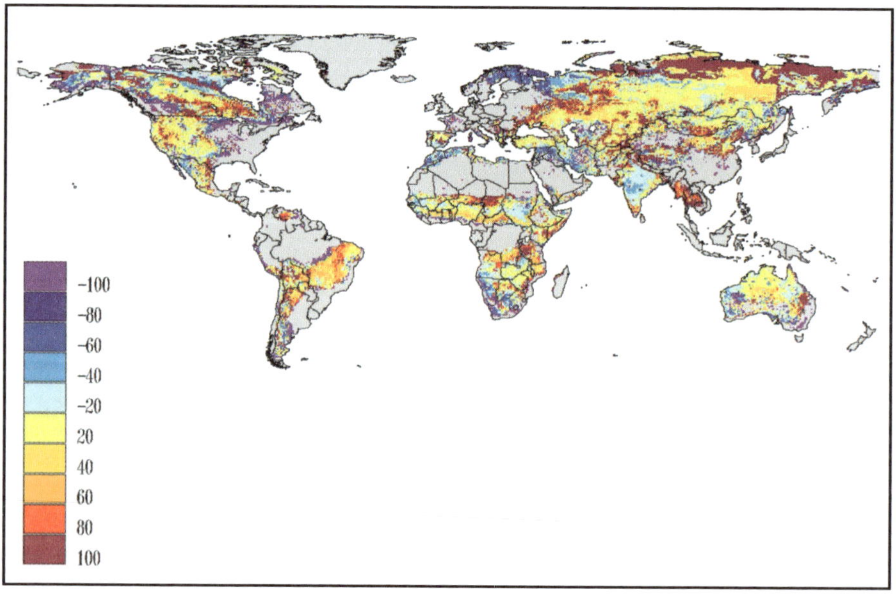

Figure 5-2. Comparison of the percent of biomass burned worldwide from 1951–2000 compared with projections for 2051–2100 (Source: Neilson 2007).

50 million t[onnes] of carbon monoxide and other radiatively active trace gases and aerosol particles" (Global Fire Monitoring Center 2003, 8). The Global Fire Monitoring Center predicts that wildfires in Russia may consume 15 million to 20 million hectares (37 million to 49 million acres) per year during the next decade, and that the areas affected by wildfires in the Russian Federation "will increase by at least 50 percent or double over the next three decades" (Global Fire Monitoring Center 2003, 11).

Fuel Treatments

In 2000, in response to catastrophic wildfires, President Clinton convened a team of experts to craft a plan to focus federal efforts in preparing for and responding to wildfires. To protect communities and valued resources, the National Fire Plan recommended the reduction of hazardous fuels, which contribute to extreme fire behavior (SAF 2002). Federal agencies have emphasized fuel treatments under the plan, treating nearly 20 million acres from 2000 to 2006; more than half of these treatments were in the wildland-urban interface (USDI and USFS 2007). Treatments are principally intended to protect communities from catastrophic wildfire losses, but also serve to retain forest habitats across broader landscapes, thus ensuring the watershed, recreational, and economic benefits of forests for future generations.

Recently, federal agencies have shifted funds from land management programs, such as timber, wildlife, and recreation management, to wildfire suppression and hazardous fuel reduction. Federal agencies, especially the US Forest Service, "are compelled to transfer an ever-increasing amount of funds to fire suppression at the expense of other programs. In the past 18 years, the wildfire management portion of the agency's budget has gone from 13 percent to 45 percent" (McMahon 2007, 2).

Despite large fire budget increases, initial fire suppression success has remained relatively stable at around 98 percent; however, if fires escape initial efforts to contain them, large areas often burn. Climate change contributes to this challenge, as do administrative factors like reduced fire staffing, fewer elite crews trained to attack high-intensity wildfires, and inadequate resources for air attacks and logistical support. The Government Accountability Office has recognized the funding challenge: "as wildland fires become more frequent and severe as the climate changes, the costs of firefighting and rehabilitating land [increase]" (GAO 2007, 31).

Mason et al. (2006) see substantial net benefits from fuel removals despite the possibility of very high treatment costs, justifying public investments in reducing the risk of fire. Although hazardous fuel treatments can be costly because the small-diameter material is expensive to remove and has little merchantable value, the future costs of wildfire have been shown to be greater than the cost of treatments if one accounts for the many costs and benefits—not just the savings in firefighting but also the avoided fatalities, property losses, timber and wildlife habitat losses, postfire regeneration and rehabilitation costs, loss in community values, hydrological damage, and carbon emissions.

Forest thinning and fuel reduction treatments often create similar posttreatment stand structures. Forest thinning reduces competition for soil moisture and nutrients, helping trees resist attacks from insects and disease and withstand drought and weather anomalies. Thinning also removes dead trees and increases average tree diameter, providing landowners with increased revenue. The principal objective of fuel reduction treatments is to reduce "ladder fuels" that increase the potential for a wildfire to reach into the crown. Additional objectives are to reduce crown bulk density and to open up the canopy so that fuels are no longer continuous. These actions reduce the potential for wind-driven fires to carry from tree to tree (Peterson et al. 2005). Both treatments typically reduce stand densities from unnatural, overdense conditions of many hundreds of trees per acre to a fraction of that level (typically 25 to 60 trees per acre, depending on age and site conditions). A "thinning from below," or "low" thinning, can accomplish both fuel reduction and growth-and-yield objectives by removing the smaller trees in the stand. These types of thinnings are ideal for biomass and smallwood markets because they use materials that might be consumed by wildfires (and produce GHG emissions) and generate energy with biomass (rather than fossil fuels). Studies have shown that woody biomass diverted for use in a bioenergy plant can reduce carbon emissions by 90 to 99 percent compared to open burning (Western Governors' Association 2006).

Fire is an essential ecological process in most forest landscapes, often serving as the primary recycling mechanism. Used appropriately, prescribed fire can be an effective hazardous fuel reduction strategy and an ecologically sound process. However, as noted by the US Forest Service and The Nature Conservancy, "short of rekindling primordial fires, the best way now to reduce the density of our forest stands that currently support many more trees per acre than in historical times is through mechanical thinning"

(Kaufmann et al. 2005, 10). Even when mechanical methods of removal are employed, a follow-up prescribed fire is usually recommended to complete the consumption of fine fuels and residual slash.

Prescribed fire has numerous ecological benefits, including restoring native plant communities, stimulating the opening of serotinous cones, providing bare soil for seedling establishment, and reducing invasive species and competition for water and nutrients. However, significant reduction in soil carbon retention and potential carbon capture (regrowth) can occur when wildfires or prescribed burns occur in dry soil conditions. In most cases, removing some of the fuels through mechanical means prior to prescribed fire can meet the ecological objectives while also reducing emissions. Other benefits of combining mechanical and prescribed fire treatments include reducing the threat of escaped fire, allowing for better protection of desired habitat components (such as snags and downed logs), and ensuring a more precise vegetative structure and treatment result. IPCC summarized the challenges and opportunities for carbon management:

> Where fuel removal is carried out, wildfire ignitions are less likely to result—and when they happen, they will often burn at lowered severities, with reduced fuel consumption, heat production, and GHG emissions. Because fire management is an integral part of forest management, it must be viewed in connection with other management practices, including harvest and wood utilization, to evaluate its full carbon flux effect. (Sampson and Scholes 2000, 271)

Mechanical fuel reduction treatments can provide an opportunity to produce valued-added forest products (engineered lumber, pulp and paper, furniture), bioenergy, and other bio-based products. These forest products (discussed in more detail in Chapters 3 and 4) are byproducts of effective fuel treatment strategies to protect communities from wildfires, yet also provide stable, living wage jobs in rural communities.

Chapter 6

Preventing GHG Emissions through Avoided Land-Use Change

Land-use change from forests to nonforest uses releases forest-stored GHGs into the atmosphere. Globally, forestland conversions released an estimated 136 billion tonnes of carbon, or 33 percent of the total emissions between 1850 and 1998—more emissions than any other anthropogenic activity besides energy production (Watson et al. 2000). Currently, tropical deforestation releases an estimated 2.6 billion tonnes of carbon annually (Malhi and Grace 2000).

Recent land-use change trends in the United States differ from the global trends (Figure 6-1). In the United States, agricultural land is decreasing, and forestland and developed land are increasing. Developed and urban lands expanded from 73 million acres to 108 million acres from 1982 through 2003, while nonfederal forestlands grew slightly, from 402 million acres to 406 million acres (NRCS 2007). Although the afforestation of agricultural lands offset the losses from development of forestlands, future afforestation opportunities will likely decrease.

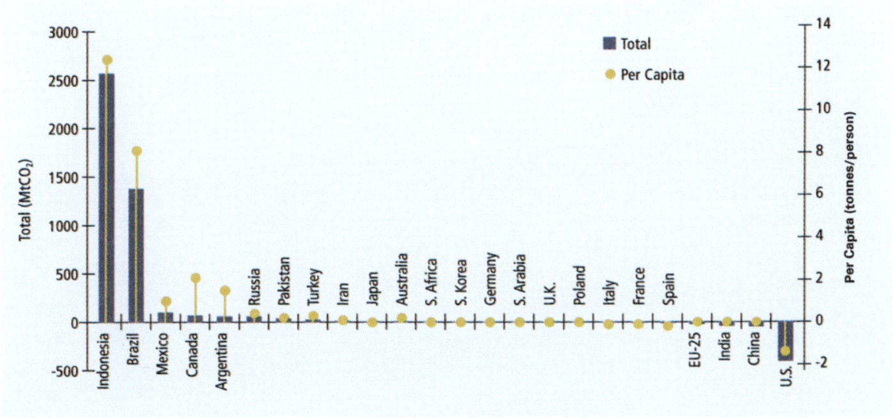

Figure 6-1. CO_2 from land-use change, total and per capita, 2000 (Source: Baumert et al. 2005, 15).

US Forests as GHG Sinks

Land uses offset approximately 14 percent of US GHG emissions in 2005 (US EPA 2007b). Forests sequestered the vast majority of those emissions (Table 6-1).

More carbon is stored in forests than in agricultural or developed land. Each year in the United States, forestlands sequester an additional 190 million tonnes of carbon in vegetation and soils (84 percent of all carbon sequestered by land use), whereas developed land sequesters only 12 percent (US EPA 2007b). Harvested biomass from forests also provides other offsets to GHG emissions when used for energy (see Chapter 4). Activities on developed lands consume more fossil fuels and produce more associated emissions than activities on forestland. Future increases in forest-based carbon sequestration (based on forest growth) depend on the availability of forestland. Loss of forestland to other uses also limits the potential positive net sequestration effects of technological advances in tree growth and silvicultural practices.

Forest conversion and land development liberate carbon from soil stocks. For example, soil cultivation releases 20 to 30 percent of the carbon stored in soils. Malhi and Grace (2000) estimate that nearly 3 billion tonnes of carbon is sequestered in the US Northeast's 117 million acres, 62 percent of which is forestland; if all those forests were developed, 400 million to 600 million tonnes of sequestered carbon would be released into the atmosphere from the soil alone (Sampson and Kamp 2007). Additional emissions would occur from the loss of the forest biomass, since a third of a tree's biomass is located in its root system (Malhi and Grace 2000). The net loss of these carbon stocks would depend on the carbon emissions and sequestration characteristics of the new land use.

Table 6-1. Net GHG emissions from land uses (MtCO_2 eq.).

Land-use category	1990	1995	2000	2001	2002	2003	2004	2005
Forestland	(598.5)	(717.5)	(638.7)	(645.7)	(688.1)	(687.0)	(697.3)	(698.7)
Cropland	(28.1)	(37.4)	(36.5)	(38)	(37.8)	(38.3)	(39.4)	(39.4)
Grassland	0.1	16.4	16.3	16.2	16.2	16.2	16.1	16.1
Settlements (urban trees)	(57.5)	(67.8)	(78.2)	(80.2)	(82.3)	(84.4)	(86.4)	(88.5)
Other (land-filled yard trimmings, food scraps)	(22.8)	(13.3)	(10.5)	(10.6)	(10.8)	(9.3)	(8.7)	(8.8)

Parentheses indicate net sequestration of GHGs (i.e., carbon sinks).
Source: US EPA 2007b.

Conversion of forestland to cropland is a case in point. Though not associated with forest biomass, nitrous oxide, which has a global warming potential (GWP) of 296, is produced naturally and emitted from soils by microbial processes. The application of nitrogen-based fertilizer to agricultural lands increases the concentration of nitrous oxide in the soil. Since 2000, the release of nitrous oxide from agricultural soil management has averaged 255 million tonnes of carbon equivalent. Within the same period, forest soils released nitrous oxide averaging 300,000 tonnes of carbon equivalent (US EPA 2006). Conversion of forestlands to agricultural lands, which is likely if energy policies favor corn-based ethanol over cellulosebased ethanol, would increase the release of nitrous oxide.

Forest soils can be a sink for methane, which has a GWP of 23. Worldwide, soils sequester 20 million to 60 million tonnes of atmospheric methane per year, equivalent to 400 million to 1,300 million tonnes of carbon (Reay et al. 2001). Soil microbes capture atmospheric methane in a process known as methane oxidation. Research has shown that forest soils are more effective than other land uses in storing methane, particularly in the well-aerated soils of temperate forests, and that the conversion of forest to other uses reduces methane oxidation. Experiments suggest that increased nitrogen inhibits the ability of the soil bacteria to oxidize methane. Methane oxidization also diminishes with increased soil moisture, such as in wetlands and peatlands, which tend to be methane sources (Bradford et al. 2001; Reay et al. 2001).

Forest vegetation also plays a complex but vital role in affecting surface temperatures. Forests tend to have a lower albedo than other land uses and thus absorb more shortwave radiation. Afforestation can therefore increase temperatures at the surface, particularly in boreal regions where there is potentially large contrast between forests and snow-covered open land (Betts 2000). Deforestation can result in increased surface albedo and, therefore, deforested areas in northern mid-latitude agricultural regions might be 1-2 C cooler in winter and spring (Betts 2001). However, deforestation in the tropics might warm the region due to changes in soil moisture, evapotranspiration and clouds (Bonan et al. 2008).*

Threats to Retaining Forestland

Increases in Land Value. Land values associated with low-density development in much of the United States have increased substantially in the past two decades while the value of land for timber production has remained

*This paragraph corrects and clarifies the relationship between forests and the albedo effect stated in Malmsheimer et al. (2008) at page 146.

stable or declined. For example, forestland in the US Southeast has been appraised for forest use at $415 per acre and for urban use at $36,216 (Alig and Plantinga 2004). Similarly, there is a significant disparity in the Pacific Northwest between the values of land for timber production ($1,000 per acre) and low-density residential development ($20,000 per acre) (Partridge and MacGregor 2007). The forestland conversion rate to urban and developed uses exceeded 1 million acres per year between 1992 and 1997, and another 23 million acres of forestland nationwide is expected to be lost by 2050 (Alig et al. 2003). This conversion would cause significant net releases of GHGs currently stored in these forests, as well as preclude future forest-based sequestration opportunities.

Landowners generally convert forestland to residential and commercial uses to capture increasing land values; however, damaging agents can also trigger conversion. When forests in the wildland-urban interface are damaged by wildfire, insects, or other disturbances, the decision to sell land for development rather than invest for long term reforestation can be attractive to landowners. Since climate change may increase the prevalence of these disturbances, forestland conversion may increase in the future.

Land values associated with agricultural crop production can reverse the recent cropland-to-forestland trend. For example, a 1999 study of Alabama Conservation Reserve Program participation found that 89 percent of acreage enrolled in tree planting was likely to remain in forests and the remainder would return to agricultural use (Onianwa and Wheelock 1999). Recent increases in agricultural crop production, especially corn and soybeans, and the development of new energy crops, such as switchgrass, may increase reversion rates to cropland. Although agricultural land is generally viewed as more environmentally desirable than developed land use, cultivation releases organic carbon into the atmosphere. Soils contain up to 60 percent of the carbon stored in temperate forests (Lal 2005). When tillage occurs on recently converted forestland, 24 to 43 percent of the soil organic carbon is emitted. In the Southeast, first-year soil-based carbon losses of 9 tonnes per acre are common but have been measured as high as 15 tonnes per acre. Soil based carbon losses decrease in subsequent years (Franzluebbers 2005).

Effects of Taxation. Property taxes and other tax policies may increase the cost of maintaining forestland and contribute to decisions that lead to forestland loss. The annual property tax on forestland is frequently the largest annual management expense in a forestry investment and results in poor financial returns followed by shifts away from forest investments (Gayer et al. 1987). Landowners are sometimes forced to sell their lands to pay the federal estate tax imposed after inheritance of forestland. For example, one study determined that 16 percent of Mississippi forestland owners who owed estate taxes sold land and/or timber to comply with the requirements (Cushing et al. 1998).

Changes in Ownership. Ownership structure affects forestland retention. Nearly two-thirds of the 620 million acres of forestland in the United States is privately owned, with 4 of every 10 forested acres being owned by "family" forest owners. In 2004, the average age of family forest owners was 60 years (Butler and Leatherberry 2004). Inheritance patterns and laws will likely increase the number of owners controlling smaller and smaller tracts. There is little evidence that the new generation of landowners will pursue commercial forest management as a primary objective. Furthermore, the logistics of implementing forest management practices become more difficult as forest tract size decreases. Even where reforestation vendors and timber harvesting companies have the ability to operate on small tracts, the high costs of these operations are often prohibitive and further encourage the abandonment of forestland management (Cubbage et al. 1989). Nevertheless, some future owners of small tracts may pursue forest management to support other goals, such as wildlife habitat improvement. Encouragement and assistance to help these small landowners pursue forestry on their lands can help maintain sequestration and storage of carbon and other GHGs.

The structure of corporate forestland ownership is also changing. Most vertically integrated forest products companies within the United States have sold large portions of their forestland holdings within the past 15 years. Within the South alone, more than 18.4 million acres of industrial forestland was sold between 1996 and 2005 (Clutter et al. 2005). The majority of this industrial forestland has been sold to timber investment management organizations (TIMOs) and real estate investment trusts (REITs), which usually pursue forest uses under a 10- to 15- year planning horizon. The financial objectives of some financial organizations could increase pressures for land-use change to development (Clutter et al. 2005).

Tools for Forest Retention

Forestland retention can occur in various ways—through public ownership of forests, higher values of forest products grown in private forests, land-use planning and related regulations on private forestlands, monetary incentives to capture the values of ecological services, and conservation easements, whether alone or as a part of other value programs.

Public Forest Ownership. Publicly owned forestland in the National Forest System and other federal and state ownerships is the least likely to be converted to other uses. Existing public forests sequester an estimated 40 million tonnes of carbon each year (Smith and Heath 2004). Efforts by states to purchase private forests continue in some areas, such as Washington State. However, concerns have been raised about the loss of the economic development values of forest production areas, diminished private ownership tax base values, and the need for government funding in other areas. This means efforts to prevent GHG releases from forestland conversion must focus primarily on retaining privately owned forestlands.

Income from Forest Products. The objectives of the private individuals and organizations that own forests range from timber production to recreational uses. Financial returns associated with forestland ownership are important to the forest industry and to TIMOs and REITs (Clutter et al. 2005). A National Woodland Owners survey indicates that approximately 30 percent of US family forestlands are owned by those who consider timber production very important or important. This number increases to 41 percent in the South, where the majority of forestland is privately owned (Butler and Leatherberry 2004). Maintaining or increasing the income potential from forest products provides incentives for forestland retention by these owners. Recent globalization and wood product substitution trends have reduced the income potential from traditional US forest-grown products (Haynes et al. 2007). However, recent advancements in developing new products, such as cellulosic ethanol and engineered wood products, may add value to working forests. Bioenergy-related products can be produced from portions of trees that have been traditionally considered nonmerchantable, as well as from the merchantable portions of trees. Sustainable utilization of working forests for a combination of wood products, including bioenergy, can improve forest landowners' returns on their land, bolster continued interest in forest management, thwart conversion to other uses, and prevent potential carbon emissions.

Land-Use Policy and Planning. Land-use planning and associated regulations have been used on large and small scales to restrict development and prevent the conversion of forestlands. Oregon's land-use planning program uses a regulatory approach to retain forest and agricultural lands. Cathcart et al. (2007) estimated that Oregon's program will prevent the conversion of 204,688 acres of forestland between 2004 and 2024. However, land-use regulation can restrict forest landowners' management options and may increase forestland conversion (Mortimer et al. 2006; Prisley et al. 2006).

Local governments have developed transfer of development right (TDR) systems to protect forestland and farmland near the wildland-urban interface (Daniels and Lapping 2005). TDRs are usually implemented through land-use planning and allow higher-than-usual-density development in certain areas in exchange for the developer's purchasing development rights from owners of nearby forests and farms (Daniels 1991).

Value of Ecological Services. Forests provide an array of ecological services, such as purifying air and water, protecting soil, and providing habitat for wildlife. These services have long been recognized and have been both regulated and subsidized by federal, state, and local governments. However, only recently have projects been developed to capture the real value of these services for private landowners. For example, an Environmental Protection Agency program allows commercial and residential developers who destroy wetlands to pay forestland owners to create and maintain wetland forests and forest riparian areas (US EPA 1990).

As Chapter 8 details, some markets for forest carbon offset projects provide landowners the ability to "capture" the ecological value that their lands provide by sequestering GHGs. These markets may provide the additional income that encourage private landowners to retain forests.

Conservation Easements. Conservation easements prevent the future development of private lands by imposing limitations on land uses and development rights (Sauer 2002). Land under conservation easements in the United States more than doubled, from 2.6 million acres to 6.2 million acres between 2000 and 2005 (Alvarez 2007). Conservation easements provide landowners tax benefits and allow landowners and easement holders to tailor development restrictions to meet the needs of each situation. The easements are typically established in perpetuity and are not easily changed after initiated. Working-forest conservation easements and the accompanying management plans generally allow for the management of forests for a variety of uses (timber, recreation, wildlife habitat) while preventing commercial and residential uses (Mortimer et al. 2007). Other conservation easements are designed to allow the ecosystem to change naturally over time with little or no vegetation management. Although conservation easements are voluntary legal mechanisms, they may be required as a condition of participation in other conservation programs, such as the trading of development rights and environmental mitigation programs.

Chapter 7

Reducing Atmospheric GHGs through Sequestration

Previous chapters have evaluated the roles of forests and forest products in *preventing* GHG emissions through wood substitution, biomass substitution, modification of wildfire behavior, and avoided land-use change. This chapter considers the role of forests and forest products in *reducing* GHG emissions. Among all possible options for reducing or mitigating GHG emissions, forests are unique in that they contribute to both goals while simultaneously providing essential environmental and social benefits, including clean water, wildlife habitat, recreation, forest products, and other values and uses.

Forest Carbon Pools

As the most efficient natural land-based carbon sink, forests play an important role in global carbon cycling. The world's forests cover 4,100 million hectares (Mha) and contain 80 percent of all above-ground carbon (Dixon et al. 1994). The greatest threat to forests is land-use change and deforestation in the tropics, which contribute about 18 percent of global greenhouse gas emissions (Stern et al. 2006). Consequently, forests are critical to stabilizing carbon dioxide and oxygen in Earth's atmosphere.

Globally, forest vegetation and soils contain about 1,146,000 million tonnes of carbon, with approximately 37 percent of this carbon in low-latitude forests, 14 percent in mid latitudes, and 49 percent at high latitudes (Dixon et al. 1994). The greatest changes in forest sequestration and storage over time have been due to changes in land use and land cover, particularly from forest to agriculture (Caspersen et al. 2000; Bolstad and Vose 2005). More recently, changes are due to conversion from forest to urban development, dams, highways, and other infrastructure.

Forestland in the United States covers 302.3 Mha (33 percent) of the land base. These forests contain 71,000 million tonnes of carbon (Mtc), with about 35 percent in living biomass, 51 percent in the soil, and 13 percent in dead material including the forest floor (Heath et al. 2003). The average rate of sequestration from 1953 to 1997, not including wood products, is estimated at 155 MtC/yr (Heath, Smith et al. 2003). A similar estimate from direct measures in 28 eastern forests during the late 1980s to early 1990s indicated a net uptake of 170 MtC/yr above ground (Holland et al. 1999).

Productive, nonreserved forestland (timberland) in the United States constitutes 204 Mha and is commonly considered the forest base potentially available for management. The average rate of carbon uptake on timberland is approximately 0.53 tC/ha/yr, with a potential uptake capacity (estimated by IPCC 2000) of 108.1 tC/ha (Kimble et al. 2003).

Because the area of US forests is so vast, even small increases in carbon sequestration and storage per hectare add up to substantial quantities. Private forestland holds 63 percent of total forest carbon, indicating the importance of private lands in policies or incentives aimed at sequestering carbon. In western forests, most carbon per unit area is in the hemlock–Sitka spruce type, which has 353.6 tC/ha; chaparral has 105.6 tC/ha. In eastern forests, aspen-birch has 309 tC/ha, and loblolly-shortleaf pine carries 163 tC/ha (Heath, Smith et al. 2003).

Urban forests are increasingly being recognized as important carbon sinks; they cover about 28 Mha, with tree cover averaging 27 percent (Birdsey and Lewis 2003; Kimble et al. 2003). This tree cover qualifies them as "forestland," which is often defined as cover exceeding 10 percent. Nowak and Crane (2002) estimate that urban trees, which cover 3.5 percent of the US land base, store 700 MtC with an annual sequestration rate of 22.8 MtC/yr. The potential for expanding the cover and extent of urban forests for both direct and indirect benefits on mitigating climate change makes them increasingly important and potentially cost-effective in sequestering and storing carbon (McHale et al. 2007).

Typically, forest soils contain a high proportion of carbon, and management practices are consequently very important in their potential effects on carbon storage. Within forest biomes as a whole, 68 percent of the carbon is in the soil, but the proportion is 50 percent in tropical forests, 63 percent in temperate forests, and 84 percent in boreal forests (Kimble et al. 2003). In southern Appalachia, Bolstad and Vose (2005) estimated the average allocation of carbon in above-ground biomass at 37 percent, mineral soil 44 percent, coarse roots 10 percent, surface litter 8 percent, and fine roots 1 percent; percentages varied depending upon the forest system. The potential net carbon sequestration in forest soils is 48.9 to 185.8 MtC/yr, with an average of 105.9 MtC/yr (Heath, Kimble et al. 2003). Immediately after harvesting, carbon in soils increases, then declines below initial values for about a decade, and ultimately increases (Heath and Smith 2000). Given the high proportion of carbon in forest soils, management of forest ecosystems should limit exposure and potential for increased soil temperature, which increases rates of decomposition, soil respiration, and erosion (Birdsey et al. 2006).

Forest CO_2 Uptake and Sequestration. In the process of photosynthesis, trees take up CO_2 from the air and, in the presence of light, water, and nutrients, manufacture carbohydrates that are used for metabolism and growth of both above- and below ground organs. Concurrently with taking up CO_2, trees utilize some carbohydrates in metabolism and give off CO_2 in respiration. Consequently, in evaluating the capacity of trees and forests to sequester and store carbon, the important metric is net carbon uptake and storage.

Because the chemical reactions of respiration are temperature driven, increases in air temperature critically affect net uptake and storage of carbon. Studies on Douglas fir and pine trees in Washington and California have shown that net CO_2 uptake is markedly lower in midday under conditions of summer stress, when temperatures are high and water content in both air and soil is low (Helms 1965). With climate change–induced higher temperatures, environmental stress is likely to increase. This will lower the capacity of plants to have positive net gains in carbon uptake, which could contribute to changes in forest type boundaries. The trend is offset to some extent by a general rise in worldwide forest productivity due to CO_2 fertilization and nitrogen deposition—both, ironically, products of anthropogenic atmospheric pollution. For example, conifer plantations in northern Britain are reportedly growing 20 to 40 percent faster than in the 1930s because of increased nitrogen deposition, atmospheric CO_2, and temperature (Cannell et al. 1998).

Net rates of CO_2 uptake by broad-leaf trees are commonly greater than those of conifers, but because hardwoods are generally deciduous while conifers are commonly evergreen, the overall capacity for carbon sequestration can be similar. There may be potential to sequester or store additional carbon in complex stand structures with mixed species compositions or several age classes due to complimentary resource use or facilitative improvement in nutrition (Kelty, 2006).*

*This clarifies a reference to mixed species, mixed-age stands in Malmsheimer et al. (2008) at page 149.

The capacity of stands to sequester carbon is a function of the productivity of the site and the potential size of the various pools, including soil, litter, down woody material, standing dead wood, live stems, branches, and foliage. In part, this is related to the capacity of stands to grow leaf area: the more leaves, the greater the stand capacity for photosynthesis and biomass production, but also the greater loss of CO_2 in respiration. Other stand dynamics that can influence sequestration capacity include age class distribution and shade tolerance. In the long run, stands of shade-tolerant species growing on high-quality sites typically have more leaf area, grow more wood, and sequester more carbon than stands of shade-intolerant species. On similar sites, stands of intolerant species initially have higher rates of wood production and carbon sequestration, which culminate earlier but do not grow as much wood, overall, as shade-tolerant species.

The rate of CO_2 uptake by trees and stands is primarily a function of species, site quality, temperature, and availability of water and nutrients. Young trees and young stands have higher rates of carbon sequestration but lower levels of total amount stored; older trees and older stands have lower rates of net uptake because, as trees age, mortality and respiration are higher. However, older stands have higher carbon storage, providing carbon is not lost to insect depredations or wildfire.

Figure 7-1 illustrates two important principles. First, young trees, and fully stocked stands of young trees, have high rates of net carbon uptake that culminate earlier for rapidly growing shade-intolerant pines than for less rapidly growing, more shade-tolerant trees, which are initially slower growing but culminate growth later and sequester more carbon overall (Figure 7-1a). Thus management practices using very short rotations of trees such as poplars and eucalypts are appropriate for intensive biomass production. Second is a general relationship involving long rotations starting from bare ground: the total amount of carbon accumulated in a given stand increases over time and reaches a plateau, after which net carbon accumulation remains relatively constant as net CO_2 uptake tends to zero because of increases in stand respiration, mortality, and decay (Figure 7-1b). Indeed, the first State of the Carbon Cycle Report acknowledges that carbon absorption by vegetation, primarily in the form of forest growth, is expected to decline over time because maturing forests grow more slowly, take up less carbon dioxide from the atmosphere, and might become carbon neutral (King et al. 2007). The report suggests that older forests could become a net carbon source

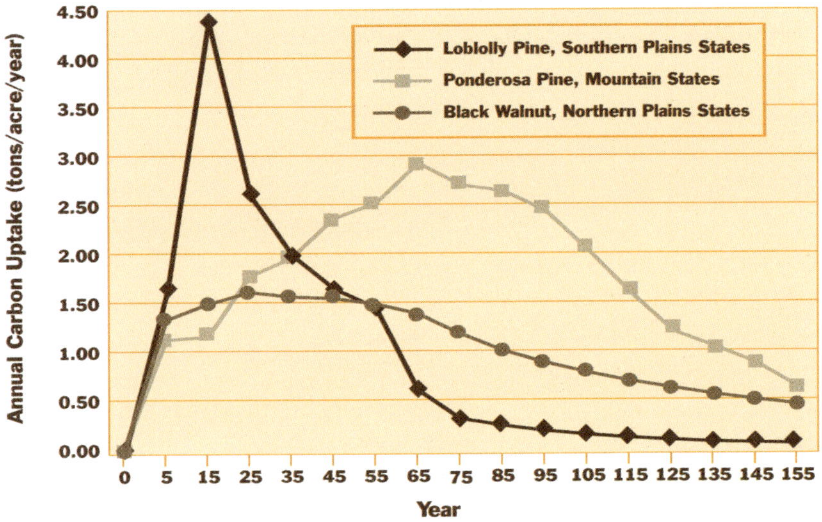

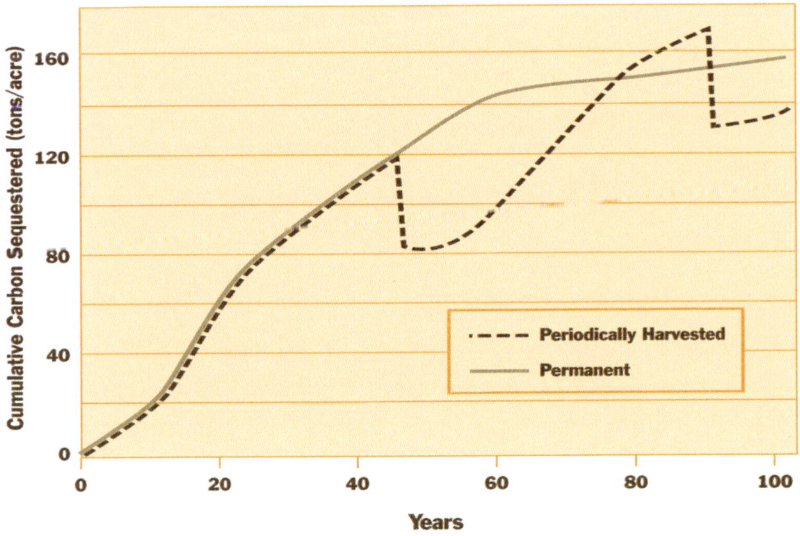

Figure 7-1. (a)(top image) Carbon sequestration rates. (b) (bottom image) Carbon accumulation rates for loblolly pine (Source: Richards et al. 1993).

because of emissions from wildfires. Figure 7-1b also shows the effect of two thinnings on carbon accumulation. In particular, after thinning, between stand ages 45 and 90 years for loblolly pine, the rate of carbon accumulation reverts to the level for stand ages 20 to 45 years. These general relationships are similar to those governing the familiar relationships between periodic and mean annual wood increment.

Carbon Release from Forests. Forests also release carbon and can become net sources of carbon to the atmosphere, particularly after a disturbance or in newly regenerated stands when soils are exposed during harvesting and site preparation. After disturbance, heterotrophic soil respiration is greatest in young forests and declines as forests age. Pregitzer and Euskirchen (2004) reported that mean temperate net ecosystem productivity in forests aged 0–10, 11–30, 31–70, 71–120, and 121–200 years was −1.9, 4.5, 2.4, 1.9, and 1.7 MgC/ha/yr, respectively. As forests become older, the amount of carbon released through respiration and decay can exceed that taken up in photosynthesis, and the total accumulated carbon levels off. This situation becomes more likely as stands grow overly dense and lose vigor, and it will become more probable in areas where climate change causes higher temperatures. However, as maturing forests become less productive, they may continue to accumulate carbon in coarse woody debris, the forest floor, and the soil.

Wildfires are the greatest cause of carbon release. In 2006, 96,385 wildfires burned 3,997,467 ha in the United States. Although 83 percent were human-caused, aggressive fire suppression policies over past decades and other factors have resulted in greatly increased fire hazard conditions that tend to make wildfires catastrophic and stand-replacing. From 1997 to 2006, 24,122,967 ha burned (National Interagency Fire Center 2007). The amount of carbon released by wildfires is difficult to estimate because of the great variability in fire intensity and fuel loads. It is estimated that every dry ton of forest biomass burned releases roughly 1.3 to 1.5 tonnes of CO_2, 0.05 to 0.18 tonnes of carbon monoxide, and 0.003 to 0.01 tonnes of methane (Sampson 2004). Average emissions might be 29 tonnes of CO_2 equivalent per hectare (Sampson 2004). Therefore, the amount of greenhouse gases emitted in 2006 through wildfires could be 128 $MtCO_2$.

Climate change–induced increases in wildfire occurrence and intensity will increase the tendency for forests to become a source rather than a sink for carbon (Dale et al. 2001; Nitschke and Innes 2006; Westerling et al. 2006). Changes in the fire regime could even overshadow the direct effects of climate change on species distribution and migration (Dale et al. 2001; Nitschke and Innes 2006). Limiting the extent of wildfires through forest management would therefore contribute greatly to mitigating climate change. For example, Lippke et al. (2006) estimated that, primarily as a result of reduced forest fire emissions and increased long-lived forest production, 56 percent more carbon could be stored over a 50-year period in a managed than in an unmanaged forest in eastern Washington.

Historically, insects and disease have caused mortality on approximately 1.6 Mha/yr in the United States (Birdsey and Lewis 2003). Recent years have seen a number of large outbreaks of pine beetles and other insects that appear to be directly related to a warming climate. In 2006, the mountain pine beetle epidemic in British Columbia destroyed 9.2 Mha of lodgepole pine forests, for a cumulative effect of 14 Mha (Carrol et al. 2004; BC Ministry of Forests and Range 2007). In 2003, 1.5 Mha of pinyon pine forests in eight states of the Southwest was affected, with mortality reaching 90 percent. Tree mortality caused by insects and disease in recent years thus equals or exceeds that caused by wildfires.

Other important forest disturbances include hurricanes, ice storms, droughts, and floods. In 2005, Hurricane Katrina affected 2 Mha of forest in Mississippi, Louisiana, and Alabama, killing or severely damaging approximately 320 million large trees and releasing, over time, approximately 105 million tonnes of carbon dioxide to the atmosphere—roughly the net annual sink in US forest trees (Chambers et al. 2007). Harvesting occurs on approximately 4Mha/yr, with 62 percent being partial harvests. Interestingly, the area harvested annually in 1907 was 3.8 Mha (Birdsey and Lewis 2003).

In general, forests may be either carbon sinks or sources, depending on their age and health. Unmanaged, older forests can become net carbon sources, especially if probable losses due to wildfires are included (Oneil et al. 2007). Because of the variable conditions of US forests, particularly overstocking on federal lands, forest management has substantial opportunities to both enhance sequestration and reduce carbon emissions, particularly by reducing carbon lost because of wildfires, insect and diseases, and avoided conversion of forests to other land uses.

Enhancing Storage and Reducing Emissions

Forests of all ages and types have remarkable capacity to sequester and store carbon. Enhancement of this capacity depends on ensuring full stocking, maintaining health, and reducing losses due to tree mortality, wildfires, insect, and disease. Addressing each of these issues requires management that controls stand density by prudent tree removal; this provides society with renewable products, including lumber, engineered composites, paper, and energy, even as the stand continues to sequester carbon. Above all, enhancing the role of forests in reducing GHGs requires keeping forests as forests, avoiding conversion to other land uses, increasing the forestland base through afforestation, restoring degraded lands, and increasing tree density on understocked areas.

The Western Forestry Leadership Coalition (2007) suggests that two active forest management approaches should be considered to enable forests to provide ecological, social, and economic benefits to society in the face of the environmental stress associated with climate change. The first approach is adaptation, which involves positioning forests to become more healthy, resistant, and resilient. The second is mitigation, in which forests and forest products are used to sequester carbon, provide renewable energy through biomass, and avoid carbon losses due to fire, mortality, and conversion. On any given area of forestland, adaptation and mitigation objectives at the same time could be either complementary or incompatible. A complementary situation would occur where activities to maintain healthy, resilient forests also reduced the risk of uncharacteristically severe wildfire, CO_2 emissions, and damage to watersheds, and where the byproducts of such activities are used to offset fossil fuel burning. Incompatible competition could occur, for example, on some parts of national forests, where the objectives of sequestering high levels of carbon may conflict with adaptation needs that require reducing carbon stocks.

Adaptation. As described in Chapter 2, climate change will likely create stress on forest systems, changing competitive relationships among species and altering the tendencies for species to be more or less successful in a given locality. In general, species are expected to move northward in latitude and upward in elevation, although there will likely be opportunistic expansions and contractions of species and communities as habitat suitability changes. Scientists suggest that existing biological communities will change as individual species move in response to changing climatic conditions and chance events. Thus, existing communities are likely to disassemble, species by species, and then reassemble, perhaps into communities or "novel ecosystems" that have no analog today (Hobbs et al. 2006). This makes predicting future plant associations exceedingly difficult.

An important question is whether management can help forest systems adapt to new environmental conditions. Can management protect, enhance, modify, or adapt to changing ecosystem values? Because past experience may no longer be a valid basis for management planning (Perschel et al. 2007; Millar et al. 2007), the first task is anticipating what kinds of changes can be used as a basis for informed decision making. In particular, Breshears et al. (2005) ask, can we identify what triggers ecosystem change and how well can we judge the extent of change? It is perhaps especially important to identify the potential response of overstory, or "keystone," species—those that will rapidly alter ecosystem type if they lose vigor or die (Breshears et al. 2005). By the end of the century, the climate of 55 percent of western US landscapes may be incompatible with today's vegetation (Rehfeldt et al. 2006). Therefore, predicting the composition and distribution of future plant communities from contemporary climate profiles in large, heterogeneous physiographic regions may be impossibly complex (Rehfeldt et al. 2006).

Already, past protracted droughts and water stress have triggered large-scale dieoffs and landscape changes. In the Southwest, massive outbreaks of bark beetle infestations have occurred in ponderosa pine and pinyon pine. Not only are these accompanied by possible shifts in forest ecotones, but there are other ramifications as well, including potential runoff and erosion, effects on associated wildlife, changed competitive relations of understory species, and altered dynamics of carbon sequestration and storage. Similarly, changed climate, particularly warmer winters, appears to be responsible for triggering the current epidemic outbreaks of mountain pine beetle in the lodgepole forests of British Columbia and Colorado.

Consideration of how management might address changed climate–ecosystem relations focuses attention on modeling. However, land managers should use model results and generalizations regarding climate change with great caution. Model projections at global and regional scales may indicate climate trends with confidence, but it is much more difficult to assess trends at the local scales important to land managers. This is particularly important in topographically complex mountainous areas, where high-quality, daily meteorological data at fine spatial scales are needed (Daly et al. 2007). It is even more difficult to assess trends in biotic responses to anticipated climate change and, with confidence, judge the likelihood of shifts in species and communities of forest biota at spatial scales consistent with local management and ownerships. Management is further complicated by the need to understand interactions among landscape fragmentation and population mobility and dynamics (Halpin 1997). Responding may incur greater risk than doing nothing (Spittlehouse and Stewart 2003).

Nevertheless, models can provide very useful guides. An example is the work of Rehfeldt et al. (2006), who modeled 35 expressions of temperature, precipitation, and their interactions in the context of plant-climate relations for the western United States. They showed that global warming should increase the abundance of montane forests and grasslands at the expense of subalpine, alpine, tundra, and arid woodlands. Important factors were the ratio of summer to annual precipitation and the summer-winter temperature differential, together with complex interactions. Rehfeldt et al. suggest that although future vegetation may retain the general characteristics of deserts, grasslands, and forests, it is commonly likely to support quite different plant associations. As climate changes, plant fitness may deteriorate, which activates evolutionary processes. Modeling efforts are becoming increasingly sophisticated, and rapid advances are being made in predictive capacity. To better guide understanding and response to change, increased capability is needed in analysis at the landscape rather than the regional level (Rehfeldt et al. 2006). A good example is the effective use of models for the Greater Yellowstone Ecosystem, where temperature and temperature-related variables have been used to describe the distribution of whitebark pine in relation to tree line (Schrag et al. 2007). Insights into the adaptation of plants to changing conditions can also be obtained by reexamining the relative performance of species and varieties planted in seed orchards and progeny test sites, and consulting studies of range-wide comparisons.

So, how might management adapt to possible climate changes? A prudent approach is that the greater the uncertainty and risk, the greater the flexibility in setting both short-term and longer-term goals and decisions (Perschel et al. 2007; Millar et al. 2007). No single solution is likely to fit all future challenges, and it is best to mix strategies (Millar et al. 2007). Three adaptive strategies based on understanding ecological processes rather than structure and function are currently being discussed (Perschel et al. 2007; Millar et al. 2007): increasing resistance, increasing resilience, and assisting migration.

Increase resistance. Resistance is the capacity of an ecosystem to avoid or withstand disturbance, such as anticipated increased insect and disease epidemics and wildfires. Management actions would aim at forestalling damage and protecting valued resources, such as water, endangered species, wild land urban interface areas, and special forest stands. Treatments to be considered include thinning of overstocked stands, prescribed burning, removal of invasive species, and restoration of native species. Since it may not be feasible to conduct treatments at the landscape scale because of fragmented ownerships and jurisdictions, implementation of this strategy could include identifying which populations are most at risk and which areas in the landscape are more likely to be buffered against the effects of changes in climate (and thus act as refugia).

The likely benefit of this approach is that it is proactive (planned and implemented before a disturbance event) and has a high probability of being successful. A potential drawback is that the scale of the disturbance could be sufficiently large to overcome the capacity of the forest to resist its effects, with negative consequences for the forested ecosystem.

Increase resilience. Resilience is the capacity of an ecosystem to regain functioning and development after disturbance. Management actions would aim at retaining desired species even if sites become less optimal. Possible treatments include 1) promoting diversity in species and age classes when replanting or conducting other treatments after a disturbance event; 2) broadening genetic variability of seedlings when reforesting after harvesting, fires, or other disturbances; 3) supporting existing forest communities while allowing transitions to new forest types; 4) identifying and enhancing possible refugia prior to disturbance; and 5) enhancing landscape connectivity so that ecological movement can take place unimpeded across the landscape, including prevention of further forest fragmentation and restoration of ecosystem processes, such as watershed function and hydrologic processes.

Likely benefits are that management can identify and plan actions in advance of a disturbance and then implement postdisturbance treatments. Planning postdisturbance actions focuses attention on which system components are most likely to be altered when changes might come about. Potential drawbacks are that actions may be taken to restore or enhance ecosystems based on past climate and experience, whereas climate change may be driving the area toward new assemblages of species. Managers should identify the appropriate vegetation communities needed for restoration forestry in conditions of change.

Assist migration. What might be needed to enable an ecosystem to adapt to changed conditions? Management actions would seek to facilitate the transition of an ecosystem from current to new conditions. Consideration would be given to introducing different, better-adapted species, expanding genetic diversity, encouraging species mixtures, and providing refugia. This approach is highly controversial—it involves taking action based on modeling and other projections for which outcomes or expectations are highly uncertain—and is in a youthful stage of development (McLachlan et al. 2007).

However, modeling at the global, regional, and landscape levels can be combined with current species climate distribution maps to suggest where tree species populations may migrate over the next century in response to various climate change scenarios. Models can possibly be used in a decision support context informing management on how to consider the potential risks and benefits of assisting migration.

Assisted mitigation might be considered in several circumstances: 1) where, after a fire or insect or disease outbreak, planting of the original species is predicted to fail; 2) on the edge of an ecotone where new species are known to be migrating into the area in a manner that validates the climate change models for the region; 3) for rare, threatened, or endangered species that are endemic to a small area and not expected to be successful in migrating without assistance; 4) new species could be added to the mix of trees being planted if these are not expected to have negative ecological consequences; and 5) where refugia have been identified as places to plant and "store" endangered species.

Assisting migration would require the development of policies and guidelines addressing the precise conditions under which species should be moved into new areas and lay out protocols for the detailed monitoring required (McLachlan et al. 2007). Because of its controversial nature and the risk of unanticipated consequences—for example, the planted species might become an invasive in its new range, or climate change might not occur in the expected manner—this level of experimentation within forested ecosystems may not win public or scientific support.

Changes in climate already appear to be occurring. It seems prudent, therefore, that adaptive approaches to management be considered. The considerable risk and uncertainty notwithstanding, the diverse values of forest ecosystems are too high to simply do nothing. The hallmarks of future forest management should be flexibility in both short-term and long-term planning, increased use of modeling, increased monitoring to detect the occurrence and direction of change, and adaptive management.

Mitigation. Whether, in the long run, managed forests can positively affect the global carbon balance compared with leaving forests unmanaged depends on several assumptions, such as the level of forest productivity, likelihood of tree mortality, uses of wood products, and extent of product substitution. Heath and Birdsey (1993), for example, projected that a no-harvest scenario sequestered more carbon. Schlamadinger and Marland (1996) commented that reduced CO_2 emissions to the atmosphere could be attained through four mechanisms: storage of carbon in the biosphere, storage of carbon in forest products, use of biofuels to replace fossil fuel use, and the use of wood products that displace other products requiring more fossil fuel for production. These authors found, over the long run, that the amount of carbon stored in the biosphere and in forest products reached a steady state, and continuing mitigation of carbon emissions depended on the extent to which fossil fuel was displaced by bioenergy and wood products. They concluded that the net carbon balance at the end of 100 years was very similar, whether trees were harvested and used for energy and traditional forest products, or the area was reforested and forest protection strategies implemented. Marland and Schlamadinger (1999) concluded that storing carbon on site in the forest and harvesting forests for a sustained flow of forest products are not necessarily conflicting options: mitigating net emissions of carbon depends on site-specific factors, such as forest productivity and the efficiency with which harvested material is used.

The issues are complex and defy easy generalizations. For some forest conditions, it is possible that early harvesting and use of wood products, while economically viable, could result in a lower rate of carbon accumulation compared with letting the forest grow to an older age before harvesting. Alternatively, focus on managing for carbon accumulation could lead to earlier harvest for some forest growth conditions. The degree to which forest management would change carbon sequestration and storage would also be influenced by whether wood use is long- or short-lived, whether the substitution offset is high or low, and whether there is high or low energy conversion efficiency.

In several cases, managed forests have been shown to sequester more carbon and have fewer emissions than unmanaged forests (Birdsey et al. 2000; Krankina and Harmon 2006; Lippke 2007; Hoover and Stout 2007). There are five prime reasons for this: 1) managed forests consist of younger trees that have higher rates of net carbon uptake; 2) managed forests are a source of wood products that continue to store carbon (in use or in landfills) for varying periods, depending on the product; 3) the use of wood products substitutes for use of alternative materials, such as steel, brick, concrete, aluminum, and plastic, all of which are based on nonrenewable resources that require much more energy in manufacture; 4) managed forests have lower greenhouse gas emissions resulting from wildfires, insect depredations, and land conversion; and 5) offset markets are more attractive for managed forests (Skog and Nicholson 1998; Lippke 2007; Krankina and Harmon 2006; OFRI 2006). Unmanaged forests can store more carbon over their lifespan above and below ground per unit area, but as they become mature, carbon accumulation reaches a steady state. Also, given fire return intervals that range from 10 to more than 100 years, there is high probability that in time, unmanaged, dense forests face a higher risk of stand-replacing fires or insect infestations than managed forests.

The modeling of stand dynamics enables a comparison of managed and unmanaged stands in terms of carbon sequestration and storage. For simplicity, researchers developed Figures 7-2, 7-3, and 7-4 for even aged stands commencing with bare ground, but comparable diagrams could be

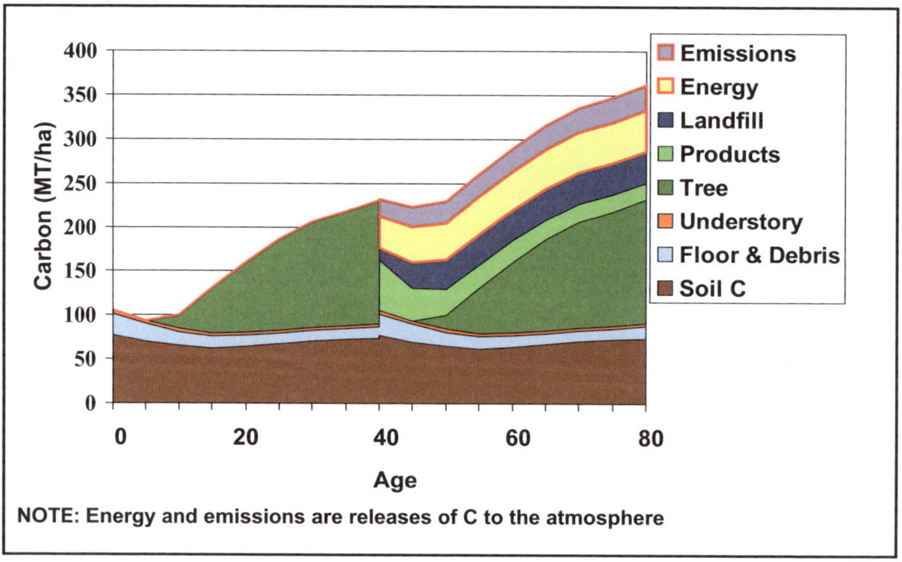

Figure 7-2. Accumulation of carbon over two 40-year rotations of loblolly pine (Source: Birdsey and Lewis 2002).

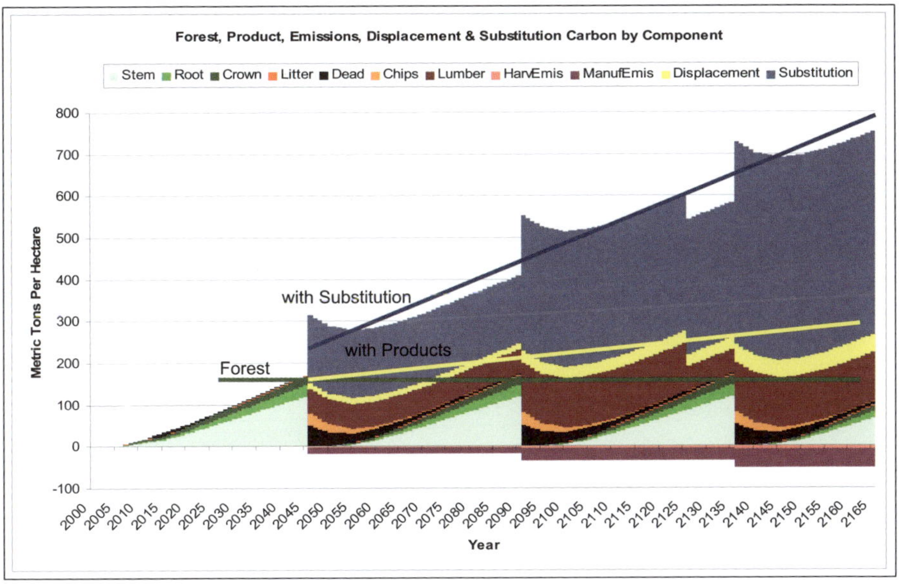

Figure 7-3. Carbon accounting over four rotations of even-aged management in Douglas-fir in western Washington (Source: Oneil et al. 2007).

prepared illustrating the growth of uneven-aged stands. Figure 7-2 shows the accumulation of carbon over two 40-year rotations of southern loblolly pine and illustrates the distribution of harvested carbon into diverse products and the decline in forest carbon stocks during the reforestation phase (Birdsey and Lewis 2002). Figure 7-3 illustrates the results of modeling the accumulation and distribution of carbon over four clear cutting rotations in western Washington (Oneil et al. 2007). Here, carbon in the forest has a stable trend line, and the carbon in product pools—net of energy used in harvesting, processing, and construction—steadily increases over time. The area in blue* shows the substantial carbon savings associated with substitution of renewable and carbon- neutral wood products for alternative, fossil fuel–intensive building products (Oneil et al. 2007).

The top diagram of Figure 7-4 illustrates the results of modeling the growth on national forests in eastern Washington and shows the forest carbon pools assuming no management, fire disturbance, or insect or disease damage (Oneil et al. 2007). The bottom diagram is a preliminary analysis incorporating the occurrence of wildfires, which because of climate change were estimated to burn 1.7 percent of the area every decade. This approximation does not

*This corrects the reference to the "gray" in Malmsheimer et al. (2008) at page 152.

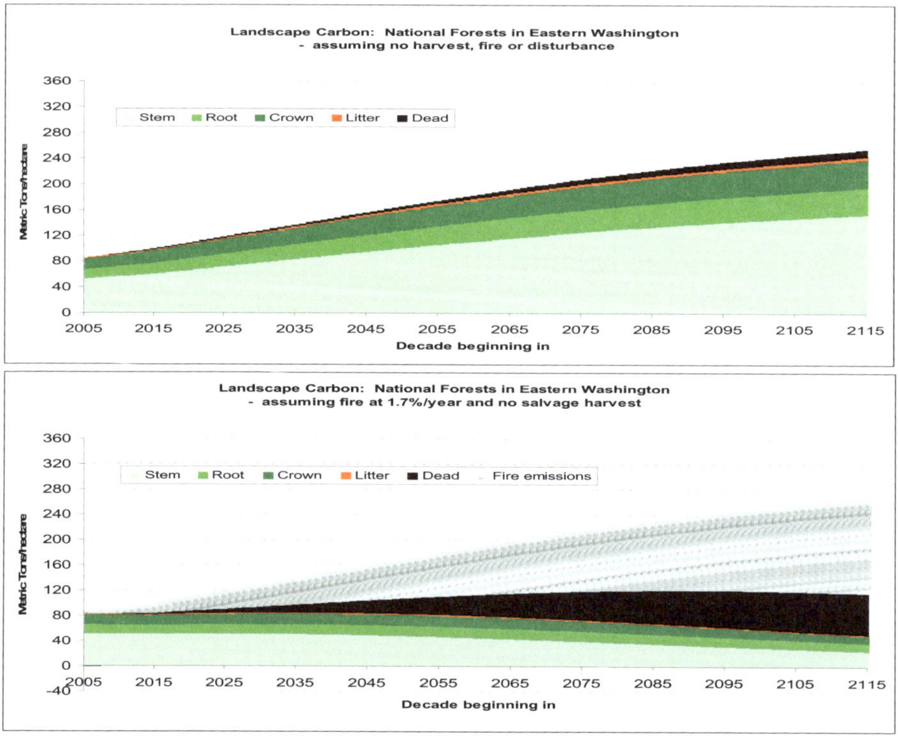

Figure 7-4. Carbon sequestration potential on national forests in eastern Washington. No disturbance compared with fire and no salvage harvest (Source: Oneil et al. 2007).

include regeneration delays and success rates, but the model outcome suggests that unmanaged national forests in Eastern Washington would likely become a carbon source rather than a carbon sink (Oneil et al. 2007).

Silvicultural Treatments That Affect Carbon. Traditional silvicultural treatments focused on wood, water, wildlife, and aesthetic values are fully amenable to being applied to enhancing carbon sequestration and reducing emissions from forest management (Helms 1996). When considering the application of alternative kinds and levels of stand or landscape treatments in the context of multiple goals and values, managers should consider it likely that attempts to enhance the output of one value will diminish the outputs of others.

Choice of management regime. One of the primary silvicultural choices foresters face is the management regime. Currently, management regimes are chosen in consideration of the economic, site, and silvical characteristics of forest stands, along with other factors. The choice of an even- or uneven-aged management regime for a forest is likely to have little effect on above-ground carbon storage over long periods of time (multiple rotations). These two broad regimes do, however, have variable carbon uptake characteristics over short time horizons, such as a rotation. By providing continuous canopy cover, uneven-aged management is likely to provide continuous carbon uptake, depending on the periodicity and intensity of partial harvest entries. In comparison, the carbon uptake under even-aged management is strongly influenced by rotation length and the length of regeneration periods when the stand has little canopy cover. Management for carbon uptake does underscore the importance of choosing the appropriate regime for each stand. Adaptive approaches to matching the appropriate silviculture with each site as a mosaic across the forest enhance overall forest productivity and carbon uptake.

Choice of species. Initially, fast-growing, shade-intolerant species have higher rates of carbon sequestration at a younger age than more shade-tolerant, slow-growing species. However, over time, shade-tolerant species are likely to have higher stand densities and leaf area and therefore higher accumulation of carbon stocks. Mixed-species and mixed age stands are likely to accumulate more carbon than single-species stands. Genetic selection, tree improvement, and biotechnology can enhance the rate of carbon uptake and storage by providing trees with higher net carbon uptake capacity. These trees are likely to have special application in growing short-rotation tree crops for bioenergy or cellulosic ethanol.

Slash disposal. Tops, needles, and branches that are residues from harvesting can be evaluated for the extent to which various treatments affect the carbon balance. Allowing this material to decay and return nutrients to the soil is a carbon-neutral process that takes several years, during which time the slash may increase the risk of wildfire. Burning the slash, although also a carbon-neutral process, immediately releases carbon, volatilized nitrogen, other greenhouse gases, and particulates into the atmosphere. Incorporating wood residues into the soil rather than burning it or leaving it to decay can increase or prolong carbon storage in the soil (Birdsey et al. 2006). Alternatively, depending on costs, this material could be used for bioenergy or the production of cellulosic ethanol. Removal of slash, however, may not be appropriate for sites with low productivity.

Site preparation. Site preparation is intended to give the desired vegetation greater access to limited resources, such as soil or water. In the context of carbon sequestration, a major consideration is limiting loss of soil carbon that follows exposure during such treatments, which may increase oxidation of soil carbon, temperature (which increases respiration of soil organisms), disturbance, and in particular soil erosion. Site preparation that incorporates wood residues into the soil can increase or prolong carbon storage in the soil (Birdsey et al. 2006).

Regeneration. Whether by natural seeding, direct seeding, planting, or some mixture of treatments, regeneration should be done promptly to minimize the time soil is exposed and the canopy is open. Prompt tree regeneration also reduces the risk that the site becomes occupied by brush, which has lower leaf area and less CO_2-sequestering capacity than trees. Early brush control has been shown to have important leverage in improving wood-growing capacity and storing carbon in both the forest and stored products (CFR 2007).

Fertilizer. Sometimes applied in planted forests and in short-rotation plantations, fertilizers increase rates of growth and leaf area production and therefore the rate of carbon uptake and sequestration. In carbon accounting, however, the source of materials used as fertilizers and the source and cost of energy used in manufacture, transportation, and application must be factored in.

Thinning and partial harvesting. Thinning and partial harvesting are techniques used in even- and uneven-aged management, respectively, to control stocking levels and stand density. The operations may be either precommercial (i.e., the thinned material is not merchantable) or commercial and are designed to improve the growth of preferred trees. The basic concept is to allocate growth and leaf area among either a greater number of small-diameter trees or a fewer number of large-diameter trees. Both treatments make openings in the canopy, and in the context of carbon storage, it is preferable to conduct light, frequent thinnings rather than heavy, infrequent thinnings. The latter create larger openings in the canopy that require a longer time to regain leaf area and capacity for carbon storage.

Rotation length. Rotation length in even-aged management influences carbon accumulation because longer rotations and larger trees increase on-site storage. (In uneven-aged management, decisions on the maximum-sized tree follow the same logic.) Longer rotations in even-aged management favor carbon accumulation because less time is taken up in reforestation and rebuilding the canopy. However, longer rotations can incur larger management costs as the value growth rates of timber fall below the expected cost of money, and delay in harvesting reduces value from other uses, including carbon storage in wood products and substitution of wood for fossil-intensive products. Longer rotations and management cycles may also involve thinnings or partial cuts to maintain forest health.

Expansion of forestland (afforestation). One of the most widely recognized forestry practices for the mitigation of climate change is the afforestation of nonforested areas to increase sequestration and storage. Because forest is the most efficient land use for carbon uptake and storage, landowners with plantable acres and degraded areas that can be restored to a productive condition have a significant opportunity to sequester carbon. Whether the land was degraded by unsustainable practices or natural events, such opportunities may provide economic incentives to turn these areas back into productive forests.

Managing for Carbon. Forest management is often categorized as even-aged versus uneven-aged approaches. Either approach may still be appropriate at the stand level; however, at the landscape level, both approaches can be used in mosaics depending on ownership objectives and stand conditions. Incorporating carbon sequestration into the suite of management objectives focuses attention on developing and maintaining high levels of leaf area because the more leaves, the more potential for photosynthesis and carbon dioxide uptake. More leaf area also increases the potential for higher respiration rates, and consequently attention must be given to net carbon uptake under the particular growing conditions.

If the goal is to immediately sequester the most carbon in the near term, shade intolerant species with high initial growth rates, grown at the highest stocking density the site will support and harvested at the culmination of mean annual increment, will sequester the most carbon in the shortest amount of time. This short rotation, even-aged forest management regime, repeated in perpetuity with succeeding rotations of shade-intolerant trees, is often said to sequester the most carbon. However, to determine the net amount of carbon sequestered, one must factor in 1) losses of soil and detritus carbon during disturbance for harvesting, site preparation, and other management activities; and 2) the carbon emissions associated with these harvesting and management activities.

If the goal is to sequester the maximum amount of carbon over a longer time frame, the best approach is to grow shade-tolerant species at the maximum stand density the site will support and implement a similar even-aged management regime, harvesting and replanting the whole stand at the culmination of mean annual increment. Shade tolerant species can be grown at a higher stand density than shade-intolerant species but have lower initial growth rates that culminate later; however, the overall amount of carbon sequestered per unit of forest area will be greater. Moreover, harvesting and site preparation activities will be less frequent and thus the associated carbon emissions will be lower.

For continuous and overall maximum sequestration, mixtures of shade-intolerant and shade-tolerant species would utilize all the photosynthetic niches in the forest canopy and forest understory while maintaining overall growth rates at a thrifty level. Uneven-aged management would use a combination of individual tree selection, crown, and understory thinning, group selection, irregular shelter wood, and other intermediate cuttings to maintain a kaleidoscope of different age classes of thrifty intolerant and tolerant trees. Again, emissions would have to be calculated for the frequent management entries, as would the combined mean annual increment for all the different species and age classes of trees, which must be discounted to an annual basis.

The important carbon sequestration metric for all three of the above approaches is the area under the mean annual increment curve, which will reveal the total amount of carbon sequestered during the management cycle. This metric can then be discounted over the time period of the management cycle to calculate the average annual carbon sequestration rate for any management scenario. Below-ground carbon sequestration in root fiber, soil, macro- and microorganisms, down woody material, and other pools must also be calculated.

If the landowner's goal is to enhance the capacity of the forest to sequester and store carbon and to reduce its likelihood of becoming a source of carbon and other GHGs in the long run, the forest should be managed. This is because, in the long run, 1) management enables the maintenance of forest health, which reduces the likelihood and severity of emissions from wildfires and insect or disease mortality; and 2) it provides products that have both short- and long-term storage capacity and can substitute for fossil fuel–based materials and sources for energy, building, and other uses. Much of the technical knowledge needed to enhance sequestration and storage is available or can be adapted from traditional practices. Knowledge gaps include the effects of management on carbon pools and the extent to which enhancing carbon reduces the outputs of other forest values and uses. There is thus a need for increased monitoring and adaptive approaches to management.

Under current economic conditions, however, carbon sequestration is not likely to be a primary management objective for most forest owners (Birdsey et al. 2006). As with any type of management, goals, costs, incentives, regulations, policy, and values will drive decisions. Carbon sequestration through forest management may, however, provide forest owners who meet requisite protocols with an additional income stream from the sale of offset credits. If realized, this additional economic return could change the economic viability of some management practices, alter the intensity with which forests are managed, and influence other management decisions. The degree to which carbon sequestration opportunities influence forest management will depend heavily on such factors as the value of carbon financial instruments, the costs of program or market participation, regulatory requirements for emission controls, market-wide recognition of offset credits from forestry projects, and opportunity costs.

Debate continues regarding the relative benefits of young, managed forests compared with older, unmanaged forests in terms of efficacy of forest carbon sequestration. But all forests, under varying levels of management or no management, can provide carbon sequestration benefits, depending on their particular condition or situation. It is important to take into account the different objectives for managing forests of varying age and the associated benefits that can accrue from older, mixed age and mixed-species forests. Indeed, there are sites of low productivity where production of timber may be so slow or uncertain that managing for forest health and fire protection could be a superior carbon sequestration strategy.

Carbon Storage in Wood Products

Harvesting reduces carbon storage in the forest both by removing organic matter and by increasing heterotrophic soil respiration (Pregitzer and Euskirchen 2004). However, much of this is offset by the carbon that is stored in forest products for varying lengths of time. The carbon in those forest products, for example, may not be released for decades. Along with the benefits of consistently high sequestration levels, it is this aspect of sustainably managed forest carbon projects that provides the maximum benefits for climate change mitigation when compared with unmanaged forests, which can suddenly release huge amounts of carbon if they burn. Forest management that includes harvesting provides increased climate change mitigation benefits over time because wood-decay CO_2 emissions from wood products is delayed (Ruddell et al. 2007). Accounting for this carbon pool is critical to accurately representing forest carbon uptake and storage on a project level. A forestry project that fails to consider it may significantly overestimate emissions from the project over time (US DOE 2007).

Until recently, carbon stored in harvested wood products (HWPs) had received little recognition in international GHG mitigation programs. In fact, the 1996 United Nations Framework Convention on Climate Change guidelines for carbon accounting for countries participating under the Kyoto Protocol considered the inputs (additions) and outputs (emissions) at the national level for the HWP carbon pool to be equal (IPCC 2006). This position was revisited in 2006 in the revised IPCC guidelines, in which HWP accounting rules for Kyoto compliant countries were presented in greater detail (IPCC 2006). The new rules facilitated a more thorough recognition of this important carbon pool, offering participating countries the option to account for carbon accumulation in this area.

In their early stages, many US climate change mitigation programs considered the harvesting of wood an immediate release of carbon. The carbon storage potential of HWPs has since become more widely acknowledged. To date, storage of HWP carbon has been recognized by some but not all domestic climate mitigation programs and registries. Although their accounting methods vary, the US Department of Energy 1605b guidelines, the Chicago Climate Exchange, the California Climate Action Registry, and the Georgia Carbon Sequestration Registry are examples of programs that now recognize this important carbon pool, though the California registry does not consider it a tradable pool at this time.

The HWP pool consists of two parts: wood in use, and wood discarded in landfills or recycled (US DOE 2006). Their interrelationships are illustrated in Figure 7-5. The delay in the release of carbon from HWPs depends on the manner in which the harvested wood is used. For example, carbon may be stored for decades in sawn lumber used in housing construction, but

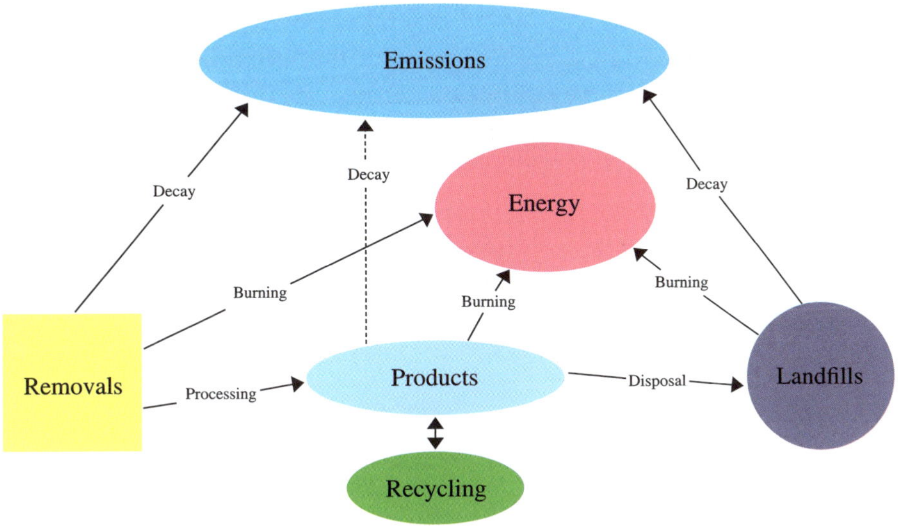

Figure 7-5. Harvested wood products pool (Source: Heath et al. 1996).

wood harvested for the production of paper may store carbon for only one to five years. Accounting approaches of the current US carbon programs vary somewhat, but most consider six basic categories of harvested wood in use: waste wood, wood used to produce energy, solid wood (lumber), composite wood products, paper products, and nonstructural panels. Each wood category has its own specific rate of decay or release to the atmosphere. One example of depreciation or half-life values for various end uses of wood products is provided in Table 7-1, which illustrates the variable decay rates specified in the US Department of Energy 1605b rules.

The accounting methods for HWPs in use fall into two main techniques. The first approach is to track, over time, the decay of materials stored in wood products and account for the specific emissions in the year in which they occur. Under this method, each harvest year is depreciated individually over a project's lifespan in accordance with the proportion of wood product types generated from the harvests. In addition to the contributions made annually to the HWP pool through harvests, annual emissions for the pool are also calculated. These calculations produce the annual net contribution to or emissions from the HWP pool. If there is a positive difference between a specific year and the previous year's HWP levels, a positive sequestration result is realized. If the result is negative, then the HWP pool

Table 7-1. Half-life for products by end use.

End use or product	Half-life (years)
New residential construction	
Single-family homes	100
Multifamily homes	70
Mobile homes	12
Residential upkeep and improvement	30
New nonresidential construction	
All except railroads	67
Railroad ties	12
Railcar repair	12
Manufacturing	
Household furniture	30
Commercial furniture	30
Other products	12
Shipping	
Wooden containers	6
Pallets	6
Dunnage	6
Other uses for lumber and panels	12
Solid wood exports	12
Paper	2.6

(Source: US DOE 2006).

has experienced net emissions and that amount would be deducted from total reported sequestration for that year. The benefits of this approach are largely in maximizing positive results over shorter project life spans and in more project-specific accounting. There are also potential drawbacks to this approach. Over longer time frames, emissions from the HWP pool could exceed total additions, resulting in carbon deficits. Also, this accounting system is somewhat complex.

The second HWP accounting method uses established depreciation tables to calculate the quantity of carbon remaining in harvested wood (also by product class) after 100 years. Based on standard decay equations, this 100-year rule allows project owners to annually retain the net carbon credits represented by the carbon estimate for their harvested wood products. The approach is much simpler and does not create net negative flows of carbon over the project lifespan. Drawbacks include fewer project-specific calculations and potentially very conservative estimates of carbon storage in the HWP pool.

If the wood product is transferred to a landfill, the time frame for the ultimate release of its carbon into the atmosphere may be even longer. To illustrate, carbon may be stored in a paper product five years after harvest, then in a landfill for 10 years, and decomposed as emissions after yet another decade or two. In accounting for carbon storage in landfills, the current US registries are even more variable. Although accounting rules for this aspect of carbon storage currently exist, this part of the pool is less uniformly recognized by domestic carbon programs than carbon stored in wood products in use. One reason involves concerns over ownership of the carbon stored in landfills, and thus who can claim credit for the carbon sequestered.

The climate change benefits of wood products are twofold: the true value lies in the combination of long-term carbon storage with substitution for other materials with higher emissions. Although some carbon accounting systems are beginning to recognize the importance of the carbon stored in wood products, fewer incorporate the system boundaries that recognize the importance of the way wood is used. Because wood can substitute for other, more fossil fuel–intensive products, the reductions in carbon emissions to the atmosphere are comparatively larger than even the benefit of the carbon stored in wood products. Research both in the United States and internationally (Borjesson and Gustavsson 1999; Buchanan and Levine 1999; Lippke et al. 2004; Lippke and Edmonds 2006; Perez- Garcia et al. 2005; Sathre 2007; Valsta et al. 2008) has suggested that this effect—the displacement of fossil fuel sources—could make wood products the most important carbon pool of all.

Chapter 8
Markets for Forest Carbon Offset Projects

Historically, command-and-control regulation has been the approach to regulating emissions and discharges of pollution into the environment in the United States. The Clean Air Act of 1970 and the Clean Water Amendments of 1972 effectively equalized pollution levels across all polluters. While effective in achieving absolute reductions in pollution, these acts prescribe technology-based and performance-based standards to pollution abatement in ways that stifle innovation and discourage the development of better, lower cost technologies (Stavins 2001). Since it is not possible for regulatory oversight agencies to know the pollution abatement cost function of each polluter, uniform standards force some firms to incur a larger cost burden per unit of production for controlling pollution.

Market-based instruments encourage the desired behavior through market signals rather than through explicit directives for pollution levels or control methods (Stavins 2001). Two such climate change policy instruments include emissions trading and carbon taxes. When well designed and implemented, these instruments create incentives that alter the producer's pollution control strategy in ways that benefit the producer while meeting pollution reduction policy goals. Compared with command-and-control approaches, market-based climate change policy instruments accomplish a cost-effective allocation of pollution control burden by equalizing the marginal costs (the incremental amount spent to reduce pollution) across all entities even though the regulator does not know their individual pollution abatement cost functions. Marketbased climate change policy instruments provide economic incentives that promote innovation in the development of pollution abatement technologies because it is in the polluter's best interest to do so (Stavins 2001).

Market-Based Policy Instruments

Emissions Trading versus Carbon Taxes. In practice, the selection of a marketbased climate change policy instrument is a political decision. This decision is based on the extent to which the instrument 1) is economically effective; 2) is cost-efficient; 3) provides social equity and fairness within and across generations; and 4) is flexible enough to adapt to changing social, political, and environmental conditions (Hanley et al. 1997).

Tradable permits are utilized within regulated emissions trading programs, also known as cap-and-trade programs. Rules for cap-and-trade programs can be highly variable. In general terms, under an emissions trading program, the allowable level of pollution (cap) within a sector is determined through a political process that allocates or auctions emission allowances among the polluting entities. In theory, the polluters will choose the least-cost means to comply with the cap. Those that keep emission levels below the cap can sell their surplus emission allowances. Those that emit more than the cap must either buy surplus emission allowances from others or, if permitted, offset their excess pollution (over the cap) by purchasing emission reduction credits from offset providers. Although emissions trading is a cost-effective policy instrument, it can also create uncertainty in the total cost of compliance for the polluter. Emissions trading programs are, however, very flexible instruments and can easily adjust to changes in the cost of emitting pollutants.

Carbon taxes are charges or penalties levied on the amount of carbon dioxide that a firm generates. Under this policy instrument, the polluter will reduce emissions to the point where its marginal abatement pollution costs are equal to the carbon tax, and thus different firms control emissions at different levels. Those with high marginal abatement costs (high-cost polluters) will reduce pollution less than those with low marginal abatement costs (low-cost polluters). One drawback of carbon taxes is that the environmental outcome—the total reduction in emissions—cannot be guaranteed because the regulator cannot know the marginal pollution abatement costs for each firm. Determining the appropriate tax rate therefore becomes a major challenge for policy makers. In theory, to achieve an economically efficient level of pollution, the tax will be applied on each unit of production at a rate that equals the social costs of pollution (Perman et al. 1996).

Emission Allowances versus Emission Reduction Credits. The design of any emissions trading program includes two primary transactions: emission allowances and emission reduction credits. Emission allowances (also called allowance-based carbon transactions) are created by a regulatory cap-and trade body and are initially allocated or auctioned to the user. Emission allowance transactions are based on the entity's direct emissions. Entities must reconcile their emissions account at the end of each compliance period through direct and verified measurements to ensure compliance with their allocated or auctioned emission allowances.

Emission reduction credits (also called project-based carbon transactions) are issued to projects that can credibly demonstrate reductions in GHG emissions compared with what would have happened without the project. Forestry is one category of projects that can provide carbon dioxide emission reduction credits (capturing landfill methane, conservation tillage practices, and alternative energy are others), and several project types are eligible (Sampson et al. 2007).

- *Afforestation:* planting trees on land that has been in a nonforest land use for a number of years (the Kyoto Protocol requires 50 years; other registries and programs require 10 or 20 years).
- *Reforestation:* planting trees on land that had previously been forested but has lost forest cover and is not recovering naturally. Severely burned forests may qualify under this definition if they show no recovery after a time period.
- *Forest management:* managing a forest to protect and/or enhance carbon stocks. The entire forest estate under management should be included to prevent the possibility that the owner will report only on areas of growing forest and avoid including the areas where the forest may be in a declining condition.
- *Harvested wood products:* providing credit for harvested wood is usually connected to forest management that includes periodic harvests.
- *Forest conservation or protection:* preventing a land-use change that would destroy or degrade an existing forest, such as conversion to agricultural or development uses. This type of offset project is also known as avoided deforestation.

Emission reduction credits should be issued only after their reductions have been verified; they can then be used to offset direct carbon dioxide emissions above a firm's allocated or auctioned emission allowances. The purchase or sale of contracts for emission reduction credits typically carries higher transaction costs and risk than emission allowances. Once emission reduction credits are issued and used to offset direct emissions, they provide the same mitigation benefit in reducing or preventing GHG emissions as emission allowances (Ruddell et al. 2006).

Programs and Markets for Forest Carbon

Project-based emission reduction credits, such as those developed through forest carbon offset projects, are used to reduce rather than prevent GHG emissions. To operate efficiently and provide the market signals required for polluters to implement the lowest-cost pollution strategy, an emissions trading program must have active trading in credits. In the absence of federal regulation in the United States, registries, voluntary emissions trading programs, and voluntary carbon offset markets have developed to satisfy demand primarily created by direct emitters wanting to reduce their GHG emissions. Mandatory emissions trading programs have become well established through the Kyoto Protocol.

The Kyoto Protocol, an international treaty of the United Nations Framework Convention on Climate Change (UNFCCC), set GHG emissions limitations on its signatory countries and established mechanisms for reducing overall GHGs by at least 5 percent below 1990 levels by the end of 2012. The protocol, which took effect in February 2005, has been ratified by all industrialized countries except the United States. Until it ratifies Kyoto or passes federal laws governing carbon emissions, the United States will remain a voluntary market for trading emission allowances and reduction credits.

Another policy option is a renewable energy credit. One type of renewable energy credit is associated with the substitution of wood-based building materials for nonrenewable building materials, such as steel, plastic, concrete, and aluminum. Research by Lippke et al. (2004), Winistorfer et al. (2005), and Perez-Garcia et al. (2005) demonstrates, through life-cycle carbon dioxide modeling, that wood-based building materials have significantly lower carbon dioxide emissions per unit of production compared with nonrenewable building materials. If these credits are recognized in US energy legislation, markets may emerge that recognize the role that this substitution plays in preventing GHG emissions.

The second type of renewable energy credit involves the substitution of wood-based biofuels, such as wood waste, for fossil fuels to generate electric power for direct emitters. Evolving carbon markets (such as the Chicago Climate Exchange) provide credits to firms with direct emissions that substitute wood-based biofuels for fossil fuels. Forest biomass is one fuel type that is being recognized as eligible for such credits under developing US Senate bills in the 110th Congress (Point Carbon News 2007).

Mandatory (Regulated) Emissions Trading Programs

Kyoto Protocol. Anthropogenic changes in Earth's climate have been the focus of climate change policy since the signing of the UNFCCC at the 1992 "Earth Summit" in Rio. To date, this convention has been ratified by 191 countries, including the United States (UNFCCC 2007c). The objective of the convention was to stabilize greenhouse gas emissions "at a level that would prevent dangerous anthropogenic interference with the climate system" (UNFCCC 2007b, Art. 2).

A global carbon market has emerged as a result of the Kyoto Protocol of the UNFCCC. Article 3 of the protocol introduced concepts of GHG emissions by sources and GHG removals by sinks, but it limited the role of forestry to afforestation, reforestation, and reducing emissions during deforestation activities conducted since 1990. In November 2001, the Marrakesh Accord provided definitions for these forestry activities and considered forest management (UNFCCC 2002). To date, only afforestation and reforestation methodologies have been approved for creating emission reduction credits for Kyoto compliance purposes. Reducing emissions from deforestation and degradation in developing countries, as part of the sustainable management of forests, was acknowledged during a December 2007 meeting in Bali, where the 13th Conference of the Parties established processes to demonstrate how such reductions could be considered climate mitigation measures and be included in the second compliance period of the Kyoto Protocol, beginning in 2013.

To combat climate change, the Kyoto Protocol uses a market-based approach—emissions trading and tradable emission reduction credits for offset projects (UNFCCC 2007b)—involving two mechanisms, the Clean Development Mechanism (CDM) and Joint Implementation (JI). Both were designed to lower the overall costs of participating countries in meeting their domestic emission reduction targets while helping developing countries and countries in transition achieve their sustainable development goals (IETA 2007).

The CDM allows Annex 1 (industrialized) countries with mandated Kyoto Protocol GHG reduction targets to invest in emission reduction projects in developing ("host") countries. In theory, these projects reduce global GHGs at a lower cost than would be possible in the Annex 1 country itself. For an afforestation or reforestation project, once a project is registered (approved), implemented, and certified, the CDM executive board issues certified emission reduction (CER) credits based on the verified difference between the baseline and the actual emission reductions that can be used toward compliance targets (UNFCCC 2007a).

JI is designed to help Annex 1 countries meet their mandated Kyoto Protocol GHG reduction targets through investments in emission reduction projects in another Annex 1 country. Verified emission reductions generate emission reduction unit (ERU) credits that can be used toward compliance targets.

The Kyoto Protocol covers only afforestation and reforestation projects, and forestry CDM projects represent only 1 percent of the 2006 volume of traded emission reduction credits (Capoor and Ambrosi 2007). As of October 2007, of the approximately 810 registered CDM projects, only 12 afforestation projects had been approved, and only one had been certified through the CDM Executive Board.

European Union Emissions Trading Scheme. An event that dramatically increased global carbon dioxide trading volume was the emergence of Phase I of the European Union Emissions Trading Scheme (EU ETS), which went into effect in January 2005. The EU ETS, the largest multinational, multisector GHG trading scheme in the world, was created to assist the 25 EU countries in meeting Kyoto Protocol–mandated emission reduction targets (European Commission 2005). Forestry activities are not eligible for either CERs or ERUs, however, effectively eliminating all international investment in forest carbon offset projects through the CDM or JI mechanisms. Table 8-1 compares traded volumes and values in the EU ETS and two other carbon markets, discussed below.

Table 8-1 Traded volumes and values of carbon credits.

Year	**EU Emissions Trading Scheme**		**Chicago Climate (CCX) Exchange**		**Over-the-counter (OTC) markets**	
	Volume ($MtCO_2$ eq.)	Value (US$ millions)	Volume ($MtCO_2$ eq.)	Value (US$ millions)	Volume ($MtCO_2$ eq.)	Value (US$ millions)
2004	—	—	2.25	2.6	—	—
2005	322	8,220	1.47	2.8	—	—
2006	1,101	24,353	10.34	38.2	14.3	58.5
2007	1,600	43,879	22.90	68.7	42.1	258.4

$MtCO_2$ eq. Million tonnes (metric tons) of carbon dioxide equivalent.

Sources: for European Union, Capoor and Ambrosi 2007; for CCX, J. O'Hara, Chicago Climate Exchange, pers. comm., November 9, 2007; for OTC, K. Hamilton et al. 2008.

Regional Greenhouse Gas Initiative. The Regional Greenhouse Gas Initiative (RGGI), a 10-state program in the US Northeast for reducing GHG emissions, will be the nation's first cap-and-trade carbon program when it goes into effect in 2009. Its goal is to reduce CO_2 emissions 10 percent by 2019. Emission reduction targets are limited to large power plants—those with energy production capacity greater than 25 megawatts—that burn fossil fuels to generate more than half of their electricity. The RGGI rules allow for the use of emission reduction credits from offset projects based on market prices for those credits. The lower the price of CO_2, the fewer the emission reduction credits that can be applied against a plant's emission reduction targets. Sequestration of CO_2 from forestry projects is limited to participating in afforestation projects. However, RGGI has contracted with the Maine Forest Service to learn how other forest carbon offset project types might be included. To date, no forest offset projects have been registered with the RGGI program.

California Climate Action Registry. In 2001, California Senate Bills SB1771 and SB527 created the California Climate Action Registry (CCAR), the nation's first statewide GHG inventory registry. Like other registries, CCAR develops rules for the issuance, qualification, quantification, verification, and registration of emission allowances and emission reduction credits for forest carbon offset projects. The Global Warming Solutions Act of 2006 (AB32) mandates that the state reduce its GHG emissions to 1990 levels by 2012 across all sectors of the economy and assigns responsibility to the California Air Resources Board to implement the cap, which will likely require emissions trading. Credits for afforestation, managed forests, and forest conservation (avoided deforestation) are allowed, and offset project rules are defined by CCAR's Forest Sector Protocol (CCAR 2007). To date, credits from one forest carbon offset project have been registered and sold.

Voluntary Markets for Forest Carbon. Voluntary carbon markets are developing globally to address the increased demand to reduce GHG emissions where not otherwise required by Kyoto, RGGI, CCAR, or other regulations. The global voluntary carbon market includes over-the-counter transactions and emissions trading transactions through the Chicago Climate Exchange (K. Hamilton et al. 2007).

Chicago Climate Exchange. The Chicago Climate Exchange (CCX) is the world's first and North America's only legally binding rules-based GHG emission allowance trading system. CCX is also the only global system for emissions trading of all six greenhouse gases. Members make a voluntary but legally binding commitment to meet annual reduction targets of 6 percent below baseline emissions by 2010. Members that reduce below the targets have surplus allowances to sell or bank. Those that emit above the annual targets comply by purchasing emission reduction credit contracts, called carbon financial instruments. Table 8-1 provides traded volumes and values on CCX.

Emission allowances are issued in accordance with a member's emissions baseline and the CCX emission reduction schedule. Integrated commercial forest entities that own mills and comply with a sustainable forest management standard with third party verification have the option of claiming their forest operations as carbon stable or using an approved forest growth-and-yield model to account for the annual net change in forest carbon stocks as a part of an entity wide accounting of GHG emission allowances.

Nonmembers can also use the CCX trading platform. The forest carbon offset projects that are eligible to be registered and traded by approved aggregators or offset providers on CCX include afforestation, reforestation, sustainably managed forests, and forest conservation (avoided deforestation). The CCX forest carbon offset rules also allow for the counting of long-lived harvested wood products in use. Annual verification of net changes in carbon stocks by an approved verification body is required before emission reduction credits can be registered and traded.

Over-the-counter markets. Society's heightened awareness of global warming has led many organizations and individuals to look for ways to mitigate their own greenhouse gas emissions. Terms such as "carbon footprint" and "carbon neutral" have entered the vernacular. Many environmentally conscious organizations and individuals have sought to mitigate their personal contributions by participating in the above registries and markets, and also through other voluntary direct sales, frequently referred to as over-the-counter (OTC) transactions. OTC transactions provide a wide range of global opportunities. Large organizations can invest directly in specific mitigation projects that meet their environmental, cost, and/or GHG mitigation objectives. Individuals can mitigate on a smaller, more retail scale.

Suppliers of carbon offset projects within the OTC have generally been classified as offset project providers, developers, aggregators, wholesalers, and offset credit retailers (K. Hamilton et al. 2007; Clean Air–Cool Planet

2006). OTC suppliers are a highly fragmented group of for-profit and not-for-profit conservation and private sector organizations that allow polluters to offset their direct emissions, and retailers can sell credits to consumers who want to offset the GHG emissions of their personal activities, such as travel. Suppliers include well known organizations such as the Climate Fund, Conservation Fund, Pacific Forest Trust, New Forests, Terrapass, and The Nature Conservancy. Credits issued in OTC markets are referred to as voluntary (or verified) emission reduction (VER) credits to distinguish them from CER credits issued under a certified UNFCCC CDM project. Private corporations are the single largest buyers of emission reduction credits in OTC markets.

Currently, there are no uniform standards under which voluntary offset projects are developed and sold. The various standards that do exist typically define approved baseline methodologies and test for additionality, permanence, and leakage (discussed below). Offset projects for the OTC market apply a variety of design elements defined by either the supplier or the buyer of the credits, but this is changing. The lack of standards for OTC market transactions has led to several standards development efforts:

- Voluntary Carbon Standard, a global benchmark standard for project-based voluntary emission reductions;
- Gold Standard, a voluntary standard designed to improve the quality of CDM and JI and voluntary offset projects;
- Green-e, a voluntary certification program that sets consumer protection and environmental integrity standards for GHG reductions sold in the voluntary market; and
- Harnessing Farms and Forests, a technical guide on the implementation of offset projects developed by scientists at Duke University, Environmental Defense, and elsewhere.

These standards define rules that can be adopted by suppliers or prescribed by buyers to create transparency, primarily in the quality of clean technology project development. However, the standards may not be wholly appropriate for sequestration offset projects like forestry. Table 8-1 provides traded volumes and values on the OTC market.

CCX and OTC CO_2 demand curves and prices. Since the OTC market is voluntary and not driven by compliance requirements, demand for OTC offset project VERs and the prices paid for CO_2 are not publicly available. Two primary differences distinguish these voluntary forest carbon markets.

One is their CO_2 demand curves. Offset credits in CCX are registered as a fungible commodity—that is, they are not distinguishable from other carbon offset project credits, such as conservation tillage, alternative energy, or landfill methane projects. On the CCX trading platform, "a ton is a ton." Because demand is derived from compliance with CCX emission reduction commitments, the quality of an offset project is determined by the CCX rules, which provide consistency across the varying forest project types. In contrast, within OTC transactions, offset credits are not a fungible commodity; the rules behind them are important purchasing criteria that distinguish offset projects and enable buyers to discriminate among them.

The other difference is the way the price of CO_2 is determined. In the OTC market, project design and benefits are important criteria that determine the value of credits from forest carbon offset projects. For example, forest projects typically include design elements that provide for social and conservation co-benefits, such as improved water quality and promoting biodiversity goals. For suppliers selling credits into OTC markets, buyers discriminate among projects based on these environmental, social, or economic benefits. The demand for and the price of CO_2 are driven by the quality characteristics of the project's design and the social and conservation benefits it produces. Therefore, "a ton is *not* a ton" on OTC markets, as it is with the fungible CO_2 commodity traded on CCX.

Those two primary differences are reflected in the current prices paid for OTC and CCX forest carbon offset credits. In a recent survey of more than 70 suppliers to the voluntary carbon market, K. Hamilton et al. (2007) found that social values, additionality, environmental quality, and certification were more influential purchasing criteria than price, advertising, or convenience. Because buyers of carbon credits may be interested in an array of conservation and economic values provided by forest projects, registries and providers that offer offset credits of high quality are frequently able to generate higher prices.

Economic Factors of Forest Carbon Offset Projects

Perhaps the most significant decision that influences economic factors in the voluntary carbon market is the choice of market a forest project owner participates in—CCX or OTC. Compared with emissions reductions from clean technology, forest carbon offset projects have unique characteristics that mean higher transaction costs. Each of the multiple registries and programs in the United States has its own rules for participating—the setting of carbon baselines, the eligibility of managed forest versus afforestation and reforestation, monitoring methods, verification rules, the pools of carbon that can be registered (i.e., above ground, below ground, harvested wood products)—all of which can raise transaction costs for organizations that manage forestlands in multiple regions of the nation (Ruddell et al. 2006).

For many forest owners, participation in new environmental markets will require new investments. Most registries and programs require an initial investment and ongoing participation costs throughout a project's life. Common examples of startup costs include conducting a forest inventory to program specifications, securing third party certification to a recognized sustainable forest management standard (such as the Forest Stewardship Council, standards endorsed by the Program for the Endorsement of Forest Certification, the Sustainable Forestry Initiative, and the American Forest Foundations Standards of Sustainability), and developing new accounting mechanisms to track the annual net change in carbon stocks. Participation involves registration and trading fees, aggregation or broker fees, costs of verification, monitoring and remeasurement costs, annual reporting expenses, and possible costs of additional insurance policies.

One influential factor for forest owners is the opportunity cost associated with forest carbon offset projects. Opportunity costs can be difficult to quantify because they differ from one project or program to another.

A potentially significant opportunity cost that needs to be considered by project owners involves permanence. Many of the current registries and programs require that forest projects remain as forests for a certain length of time to ensure the permanence of any credits sold. Two mechanisms typically used to accomplish permanence are deed restrictions on land use and long-term or permanent conservation easements. Both can increase the opportunity cost of investing in or maintaining ownership of forests for climate change mitigation.

This issue is problematic for sustainably managed forests because investors, policy makers, and buyers of carbon offset projects may not fully understand how opportunity costs apply in forestry. Forest carbon offset projects must absorb the opportunity costs associated with keeping the forest intact, forgoing potential profits from development or conversion to other land uses. In the case of permanent conservation easements, the opportunity cost of forgoing land development (forever) may be enormous—a reality not currently reflected in compensation mechanisms (Ruddell et al. 2006).

Accounting for Forest Offset Projects

The standards discussed above are attempts to provide consistent rules under which all offset projects can participate (Ruddell et al. 2007; Sampson et al. 2007). Since a major purchasing criterion for offset buyers is project quality, standards create value for buyers and suppliers, as well as financial institutions and investors, but the current standards were developed primarily with clean technology projects in mind, not sequestration projects like forestry.

Through mandatory markets driven by the Kyoto Protocol, forest project participation has been restricted to afforestation. To date, only 12 afforestation projects have been approved under Kyoto, and one has been certified through UNFCCC's CDM executive board. The main reason for the paucity of sequestration projects is that they present unique accounting issues. Cathcart and Delany (2006) and Ingerson (2007) describe and discuss carbon accounting issues in detail; here, we briefly discuss additionality, baseline setting, permanence, and leakage as they apply to forestry.

Additionality and Baseline Setting. Since benefits to the environment are the goal of any emission reduction credits program, the net amount of carbon sequestered must be additional to what would have occurred without the offset project. For forest projects, additionality can be difficult to demonstrate. A carbon baseline must be established against which the net change in carbon stocks is measured so that emission reduction credits can be quantified, verified, and registered. Typically, baseline carbon values are determined through standard forestry biometric methods that include direct and statistically designed and modeled measurement techniques.

Two types of baselines used in US registries and programs are the business as usual (BAU) and base-year approaches. The BAU scenario is based on the proposition that emission reductions that would (or might) have happened in any event should not be allowed to offset industrial emissions. This scenario works well for clean technology but not for land-based sequestration practices, where natural ecosystem dynamics and unpredictable future human actions make any projection highly uncertain.

Changing forest management objectives, markets for alternative land uses, timber prices, and ecosystem service prices (e.g., the price of sequestered carbon) all contribute to a high level of inherent uncertainty when defining a baseline under the BAU scenario. No credible methods currently exist to separate the effects of management action on a forest from those of environmental conditions over time. Given the current trend of converting sustainably managed forestland and high-value forest ecosystems to other uses, such as housing, it is clear that BAU cannot be applied to forestry unless it is redefined. Unlike the baseline emissions of a direct emitter of CO_2 (a coal-fired power plant, for example), which are precisely measured and operationally controlled, forest BAU baselines cannot be defined with certainty, and under the current rules, if the BAU baseline cannot be precisely defined, the project cannot be quantified, verified, or registered.

In the base-year approach to establishing a baseline, an inventory is taken at the beginning of the project period, and a second inventory is conducted some years later, using the same inventory design. The net change in carbon stocks (of all allowable carbon pools within the forest offset project) represents the carbon sequestration in the forest for that period of time. In a sustainably managed forest, this net change in carbon stocks will include all forest management actions, such as harvesting, tree planting, and fertilizing. It will also reflect the effects on carbon stocks of natural events like weather, wildfire, and insects and disease. This carbon accounting systems thus accounts for (and verifies) the total net change (positive or negative) in carbon stocks associated with both natural events and human management.

Permanence. When forest carbon credits are used to permanently offset industrial emissions, the forest project must demonstrate permanence. Ensuring that a forest project is permanent can be difficult if not impossible, however, since some of the carbon sequestered might be released through natural events, such as wildfires and hurricanes, or through management activities, such as harvesting. Some registries and programs require that any released carbon be included in the net change calculations so that credits previously issued can be paid back; no additional credits can then be issued until the net change in carbon stocks is again positive.

The mechanisms typically used to accomplish permanence—deed restrictions on land use and long-term or permanent conservation easements—can provide protection against land-use change but have no force against catastrophic disturbances that may destroy the forest carbon stocks. If conservation easements mandate prescriptive forest management practices based on current technology or requirements like mandatory reforestation, they may create future barriers for meeting additionality requirements.

An alternative approach is to enter into short-term contracts with project owners to sequester and maintain forest carbon stocks. These contracts protect the buyer or market of carbon credits from loss during the contract period. If the forest carbon stocks are lost, the buyer or market must be reimbursed. At the end of the contract, the ultimate buyer (the polluter) is still liable for those emissions and must either cover the obligations by repurchasing forest credits that are still valid or find other sources of offsets.

Leakage. Leakage is the indirect or secondary effect that a project might have outside the boundaries of the project itself. Large projects, for example, may shift activities in unintended ways, as when an afforestation project in one location displaces an afforestation project in another area. Or a project may alter the supply and demand forces of forest product markets and consequently the total area of forestland. Several kinds of leakage are possible.

- *Internal leakage:* when the project causes activities to shift within a forest operation. For example, the carbon sequestration created in one portion of the ownership prompts the owner to carry out carbon emitting activities elsewhere.
- *External leakage:* when one forest owner's action causes other owners to change their behavior. For example, where the rules for developing forest carbon projects require sustainable forest management certification, one forest owner's actions may increase the area of certified forestry in the region. Or a forest project that halts land clearing for agriculture in one place causes farmers needing land to move and clear another forest. Or project rules require a large forest owner not to harvest, reducing supplies of lumber and prompting producers elsewhere to respond by harvesting more timber.

Whether positive or negative, leakage can be very difficult if not impossible to measure for forest offset projects. Past efforts to quantify leakage have been generally theoretical and remain hard to apply to a specific situation. There are currently very few empirical data that reliably establish leakage for all forest carbon offset project types. Murray et al. (2004) suggested establishing leakage (discount) rates that would require a leakage factor to be applied at the regional level for specific activities; all projects within that region should then factor that discount into their calculations. However, the decision to adopt this or any other methodology will be a political decision, since the validity of leakage discounts will be based on the assumptions made in the analyses for a specific forestry activity.

Although most leakage discussions consider how a project might cause other owners to increase emissions or reduce sequestration, some efforts seek to prevent the internal leakage that could occur if an owner counted carbon on rapidly growing areas while not inventorying areas that were in decline for any reason. Registries and programs tend to cover this through two approaches. The first is to require forest-wide reporting, such that all forestlands in the ownership are included in any reporting. The second is to require that the project demonstrate that it is certified as meeting the requirements of an internationally recognized sustainable forest management standard. Certification of forest carbon offset project lands provides three distinctive advantages: 1) buyers are assured that the quality of the carbon credits is high; 2) in well-functioning forest product markets, where sustainable forest management is practiced across the entire forest ownership, leakage will not be an issue; and 3) certification standards may provide the foundation for carbon accounting systems.

Current accounting systems may not adequately cover all aspects of leakage at the project level and for product use. Many greenhouse gas mitigation programs have yet to fully acknowledge leakage across all forest carbon pools.

Equivalence. Since forest carbon offset projects compete against clean technology projects in voluntary markets, forestry credits must be equivalent as climate mitigation measures. Forest carbon stock changes are typically derived from statistical sampling (direct measurement), reference tables, or models, however, and therefore the measurements will be less accurate than those for clean technology projects, whose emissions are measured with a high level of precision, using meters. Most forest project proponents have encouraged project developers to make carbon stock measurements, calculations, and projections intentionally conservative by using discounting methods. Because significant discounting can be a disincentive for offset project development, particularly at low CO_2 prices, the main challenge is establishing a policy that balances discounts and other related transaction costs with statistical precision and measurement accuracy. One idea is to discount the growth portion of forest credits to provide conservative estimates for CO_2 and thereby strengthen the additionality and permanence of a project. Insurance instruments or reserve pools can also be effectively used to accomplish similar results.

Policymakers' Task. The Kyoto Protocol and subsequent Conference of the Parties meetings have identified forest project accounting issues that are handled differently by the US registries and programs and thus affect eligibility and transaction costs for potential participants. The current definitions for forest carbon accounting principles were developed several years before forest carbon offsets were recognized by UNFCCC as a way for direct emitters of CO_2 to meet emission reduction targets. As a result, these definitions do not fully reflect the important role of sustainably managed forests as carbon sinks for climate change mitigation. The forestry community needs to rethink the accounting principles. The goal should be to ensure that offset rules are appropriate for all offset project types, including managed forests, and promote additional and long-term forest carbon sequestration benefits.

Conclusion
Opportunities and Challenges for Society, Landowners, and Foresters

Seven conclusions are apparent from the analyses presented in this book:

1. The world's forests are critically important in carbon cycling and balancing the atmosphere's carbon dioxide and oxygen stocks.
2. Forests can be net sinks or net sources of carbon, depending on age, health, and occurrence of wildfires and how they are managed.
3. Forest management and use of wood products add substantially to the capacity of forests to mitigate the effects of climate change.
4. Greenhouse gas emissions can be reduced through the substitution of biomass for fossil fuels to produce heat, electricity, and transportation fuels.
5. Avoiding forest conversion prevents the release of GHG emissions, and adding to the forestland base through afforestation and urban forests sequesters carbon.
6. Existing knowledge of forest ecology and sustainable forest management is adequate to enable forest landowners to enhance carbon sequestration if there are incentives to do so and if carbon and carbon management have value that exceeds costs.
7. How global voluntary and mandatory markets develop will play a significant role in establishing the price of carbon dioxide and thus creating the incentives to ensure that forests play a significant role in climate change mitigation.

Given those facts, society's current reluctance to embrace forest conservation and management as part of the climate change solution seems surprising. Time is of the essence. Forest management can mitigate climate change effects and, in so doing, buy time to resolve the broader question of reducing the nation's dependence on imported fossil fuels.

Opportunities, incentives, and recommendations for including carbon storage as part of the forestry solution vary markedly depending on ownership and market and nonmarket considerations. It is essential that natural resource professionals provide leadership in recognizing these opportunities and in encouraging the development of incentives that enhance forest conservation and management.

Ownership Considerations

US forests are owned by a diverse array of federal, state, industrial, nonindustrial corporate, nonindustrial family, and tribal entities. The forests themselves differ markedly in species, composition, stocking, and productivity. Each ownership manages its forests, either intensively or extensively, under different policies and regulations, and each has different goals, objectives, and incentives that determine how the land is managed. Specific opportunities to incorporate carbon storage as part of management will be highly dependent upon the particular forest and forest owner. Over arching policies, programs, and incentives to enhance carbon sequestration must recognize this diverse ownership pattern and encourage partnerships and collaboration. This will require substantial effort in technology transfer, education, and information outreach.

Private forest owners and public land managers should investigate developing opportunities for incorporating carbon storage and addressing the challenges of climate change into management objectives for their respective forest ownership type, whether the opportunities are market or nonmarket based.

Market Considerations

Private forest owners and managers must monitor the developing forest carbon sequestration markets and become familiar with the concepts of carbon pools, carbon baselines, additionality, permanence, and leakage. As the markets for forestry offsets develop, the standards associated with these concepts will become better established. Specific forest tracts within specific ownerships and operating with set objectives will have varying degrees of opportunity to market carbon offsets, based on how these standards develop. For example, a forest managed on a sawtimber rotation primarily to produce wood building products might have little opportunity to market carbon credits unless wood-frame structures are accepted as a pool for carbon storage.

It is impossible to accurately predict how a future carbon market will develop and how that market will affect forest owners. At recent traded values of CO_2 equivalents, income from carbon offset projects would not be high enough to preempt forest management practices employed to produce traditional forest products. However, this potential income would likely provide incentive to alter management practices to produce some level of traditional value combined with increased carbon sequestration. Market compensation for all ecological services, including GHG reductions, may help balance landowner income streams, thereby reducing the pressure to convert forests to other uses.

Emerging biopower and biofuels markets will likely enhance values for small-diameter materials and increase competition for traditional forest products. Although this increased revenue should benefit forest landowners, the traditional forest products industry may lose suppliers or see lower profit margins because of the new markets. Likewise, carbon trading and emission reduction credits associated with biomass power production could also benefit industries and forest owners investing in the new bioenergy industry.

Nonmarket Considerations

Management of forests is complex. It includes consideration of diverse components—soil, vegetation, wildlife habitat, water, recreation, aesthetics—as well as diverse products and values. Management involves determining what balance of revenues and outputs is desired and what costs and inputs are needed to sustain those outputs. Nonmarket forest resources, such as species diversity, clean water, enhanced fish and wildlife habitat, fire-resilient ecosystems, and scenic values, are also likely to be affected by carbon management strategies. Typically, efforts to increase the output of one forest product or value will likely decrease the outputs of others.

Carbon sequestration and storage are likely enhanced by increasing the rate of leaf area production and maintaining canopy cover. This could be accompanied by, for example, a decrease in wildlife diversity or water yields. Commercial timber production is commonly driven by value growth rate rather than volume growth rate, and thus stocking levels for timber production may be lower than if the goal were to maximize biomass production. Conversely, opportunities for pulpwood production and biomass energy will encourage higher stocking levels. If wood products are accepted as carbon pools, the mix of products from the forest may change. Possibilities for carbon management must also include consideration of spatial and temporal issues—whether one is managing stands, forests, or landscapes, and what time frames are involved. Justification for increased carbon storage will be influenced by such factors as carbon prices, policy incentives, and regulations.

The Profession

The profession of forestry is a broad field covering biological, physical, quantitative, managerial, and social components. The values, needs, and uses of forests are similarly broad. Carbon storage is a new "ecosystem service" that is being added to the management opportunities that traditionally included wood, water, wildlife, and recreation. Forest managers are already beginning to consider carbon sequestration and storage and the fate of carbon following disturbance and management treatments. In addition, foresters must consider the threats that climate change poses for forests and develop strategies to mitigate potential increases in pests, drought, severe weather events, and wildfires.

America's foresters must become informed and actively consider opportunities and effects associated with climate change so that forests and forest management can continue to both serve and enhance the welfare of society. The profession must be proactive in communicating to society the importance of growing and managing the nation's forests both for the sustainable supply of diverse values and uses and for their capacity to contribute to mitigation of the adverse effects of global climate change.

There is now agreement among many that the world is facing global climate change. It is beyond argument that forests play a decisive role in stabilizing the Earth's climate and that prudent management will enhance that role. For example, the Intergovernmental Panel on Climate Change (Nabuurs et al. 2007, 543), the preeminent international body charged with periodically assessing technical knowledge or climate change, has stated, "Forestry can make a very significant contribution to a low-cost mitigation portfolio that provides synergies with adaptation and sustainable development. However this opportunity is being lost in the current institutional context and lack of political will and has resulted in only a small portion of this potential being realized at present (high agreement, much evidence)."

The challenge is clear, the situation is urgent, and opportunities for the future are great. History has repeatedly demonstrated that the health and welfare of human society are fundamentally dependent on the health and welfare of a nation's forests. Society at large, the US Congress, state legislators, and policy analysts at international, federal, and state levels must not only appreciate this fact but also recognize that the sustainable management of forests can, to a substantial degree, mitigate the dire effects of atmospheric pollution and global climate change. The time to act is now.

Literature Cited

AINSWORTH, E.A., AND S.P. LONG. 2005. What have we learned from 15 years of free-air CO_2 enrichment (FACE)? A meta-analytic review of the responses of photosynthesis, canopy properties and plan production to rising CO_2. *New Phytologist* 165:351–372.

ALIG, R.J., AND A. PLANTINGA. 2004. Future forestland impacts from population growth and other factors that affect land values. *Journal of Forestry* 102(8):19–24.

ALIG, R.J., D. ADAMS, L. JOYCE, AND B. SOHNGEN. 2004. Climate change impacts and adaptation in forestry: Responses by trees and markets. *Choices* 19(3):1–7.

ALIG, R.J., A. PLANTINGA, S. AHN, AND J. KLINE. 2003. *Land use change involving forestry in the U.S.: 1952–1997 with projections to 2050.* PNW-GTR-587. USDA Forest Service.

ALVAREZ, M. (ED.). 2007. *The state of America's forests.* Society of American Foresters, Bethesda, MD.

AYERS, MP., AND M.J. LOMBARDERO. 2000. Assessing the consequences of global change for forest disturbance from herbivores and pathogens. *The Science of the Total Environment* 262: 263–286.

BAIN, R.L., AND R.P. OVEREND. 2002. Biomass for heat and power. *Forest Products Journal* 52(2):12–19.

BARNETT, T.P., J.C. ADAM, AND D.P. LETTENMAIER. 2005. Potential impacts of a warming climate on water availability in snow-dominated regions. *Nature* 438:303–309.

BAST, J., AND J.M. TAYLOR. 2007. *Scientific consensus on global warming: Results of an international survey of climate scientists.* 2nd ed. (revised). The Heartland Institute, Chicago, IL.

BAUMERT, K.A., T. HERZOG, AND J. PERSHING. 2005. *Navigating the numbers: Greenhouse gas data and international climate policy.* World Resources Institute, Washington, DC.

BC MINISTRY OF FORESTS AND RANGE. 2007. *Mountain pine beetle, frequently asked questions.* Available online at www.for.gov.bc.ca/hfp/ mountain_pine_beetle/faq.htm; accessed November 19, 2007.

BERGERON, Y., M. FLANNIGAN, S. GAUTHIER, A. LEDUC, AND P. LEFORT. 2004. Past, current, and future fire frequency in the Canadian boreal forest: Implications for sustainable forest management. *Ambio* 33(6):356–360.

BERGMAN, R.D. 2004. *Biomass for heat and power.* WOE-2. USDA Forest Service, Forest Products Laboratory, Madison, WI.

BERGMAN, R.D., AND J. ZERBE. 2004. *Primer on wood biomass for energy.* USDA Forest Service, State and Private Forestry Technology Marketing Unit, Madison, WI.

BETTS, R. 2000. Offset of the potential carbon sink from boreal forestation by decreases in surface albedo. *Nature* 408(6809):187–190.

BETTS, R. 2001. Biogeophysical impacts of land use on present-day climate: near-surface temperature change and radiative forcing. *Atmospheric Science Letters* 1(4):39–51.

BEZEMER, T.M., AND T.H. JONES. 1998. Plantinsect herbivore interactions in elevated atmospheric CO_2: Quantitative analyses and guild effects. *Oikos* 82(2):212–222.

BIRDSEY, R.A., AND G.M. LEWIS. 2002. *Carbon in United States forests and wood products, 1987–1997: State-by-state estimates.* Presented at the 5th State and Local Climate Change Partners' Conference, Annapolis, MD. November 22, 2002. Available online at www.fs.fed.us/ne/ global/pubs/books/epa/index.html; accessed November 19, 2007.

BIRDSEY, R.A., AND G.M. LEWIS. 2003. Current and historical trends in use, management, and disturbance of U.S. forestlands. P. 15–34 in *The potential of U.S. forest soils to sequester carbon and mitigate the greenhouse effect*, Kimble, J.M., L.S. Heath, R.A. Birdsey and R. Lal (eds.). CRC Press, Baton Rouge, FL.

BIRDSEY, R.A., R. ALIG, AND D. ADAMS. 2000. Mitigation activities in the forest sector to reduce emissions and enhance sinks of greenhouse gases. In *The impact of climate change on America's forests*, Joyce, L.A., and R. Birdsey (eds.). RMRS GTR-59. USDA Forest Service, Rocky Mountain Research Station, Fort Collins, CO.

BIRDSEY, R.A., K. PREGITZER, AND A. LUCIER. 2006. Forest carbon management in the United States 1600–2100. *The Journal of Environmental Quality* 35:1461–1469.

BJØRKE, S.Å., AND M. SEKI (EDS.). 2005. *Vital climate change graphics.* United Nations Environmental Programme.

BOLSTAD, P.V., AND J.M. VOSE. 2005. Forest and pasture forest pools and soil respiration in the Southern Appalachian Mountains. *Forest Science* 51(4):372–383.

BONAN, G.B. 2008. Forests and climate change: Forcings, feedbacks, and climate benefits of forests. *Science* 320:1444-1449.

BONNICKSEN, T.M. 2008. Greenhouse gas emissions from four California wildfires: Opportunities to prevent and reverse environmental and climate impacts. FCEM Report 2. The Forest Foundation, Auburn, CA.

BORJESSON, P., AND L. GUSTAVSSON. 1999. Greenhouse gas balances in building construction: Wood vs. concrete from life-cycle and forest land-use perspectives. *Energy Policy* 28:575–588.

BRADFORD, M.A., P. INESON, P.A.WOOKEY, AND H.M. LAPPIN-SCOTT. 2001. Role of CH4 oxidation, production and transport in forest soil CH_4 flux. *Soil Biology and Biochemistry* 33(12–13):1625–1631.

BRESHEARS, D.B., N.S. COBB, P.M. RICH, K.P. PRICE, C.D. ALLEN, R.G. BALICE, W.H. ROMME, J.H. KASTENS, M.L. FLOYD, J. BELNAP, J.J. ANDERSON, O.B. MYERS, AND C.W. MEYER. 2005. Regional vegetation die-off in response to global-change type drought. *Proceedings of the National Academy of Sciences of the United States* 102:15144–15148. Available online at www.pnas.org/cgi/content/full/102/42/15144?maxtoshow_&HITS_10& hits_10&RESULTFORMAT_&fulltext_ breshears&searchid_1&FIRSTINDEX_0& resourcetype_HWCIT; accessed February 6, 2008.

BRINKER, R.W., J. KINARD, B. RUMMER, AND B. LANFORD. 2002. *Machine rates for selected forest harvesting machines.* Circular 296 (revised). Auburn University, Auburn, AL.

BUCHANAN, A., AND S.B. LEVINE.1999. Woodbased building materials and atmospheric carbon emissions. *Environmental Science and Policy* 2:427–437.

BURNS, C.E., K.M. JOHNSTON, AND O.J. SCHMITZ. 2003. Global climate change and mammalian species diversity in U.S. National Parks. *Proceedings of the National Academy of Sciences of the United States of America* 100(20):11474–11477.

BUTLER, B.J., AND E.C. LEATHERBERRY. 2004. America's family forest owners. *Journal of Forestry* 102(7):4–14.

CALFAPIETRA, C., B. GIELEN, A.N.J. GALEMMA, M. LUKAC, P.D. ANGELIS, M.C. MOSCATELLI, R. CEULEMANS, AND G. SCARASCIA-MUGNOZZA. 2003. Free-air CO_2 enrichment (FACE) enhances biomass production in a short-rotation poplar plantation. *Tree Physiology* 23:805–814.

CALIFORNIA BIOMASS COLLABORATIVE. 2005. *Biomass in California: Challenges, opportunities, and potentials for sustainable management and development.* CEC #500-01-016.

CALIFORNIA CLIMATE ACTION REGISTRY (CCAR). 2007. *Forest Sector Protocol. Version 2.1.* Available online at www.climateregistry.org/docs/ PROTOCOLS/Forestry/Forest_Sector_Protocol_Version_2.1_Sept2007.pdf; accessed October 12, 2007.

CANNELL, M.G.R., J.H.M. THORNLEY, D.C. MOBBS, AND A.D. FRIEND. 1998. UK conifer forests may be growing faster in response to increased N deposition, atmospheric CO_2 and temperature. *Forestry* 71(4):277–296.

CAPOOR, K., AND P. AMBROSI. 2007. *State and trends of the carbon market 2006.* World Bank, Washington, DC. 45 p.

CARROL, A., S. TAYLOR, J. REGINIERE, AND L. SAFRANYIK. 2004. Effects of climate change on range expansion by the mountain pine beetle in British Colombia. In *Mountain pine beetle symposium: Challenges and solutions,* October 30–31, 2003, Kelwona, BC, Shore, T., J. Brooks and J. Stone (eds.). Information Report BC-X-399. BC Natural Resources Canada, Canadian Forest Service, Pacific Forestry Centre.

CASOLA, J.H., J.E. KAY, A.K. SNOVER, R.A. NORHEIM, AND L.C.W. BINDER. 2006. *Climate impacts on Washington's hydropower, water supply, forests, fish and agriculture.* Center for Science in the Earth System Joint Institute for the Study of the Atmosphere and Ocean, University of Washington.

CASPERSEN, J.P., S.W. PACALA, J.C. JENKINS, G.C. HURTT, AND P.R. MOORCRAFT. 2000. Contributions of land-use history to carbon accumulation in U.S. forests. *Science* 290 (5494):1148–1151.

CATHCART, J., AND M. DELANY. 2006. Carbon accounting: Determining carbon offsets from projects. P. 157–174 in *Forests, carbon and climate change: A synthesis of science findings*. Oregon Forest Resources Institute, Portland, OR.

CATHCART, J., J.D. KLINE, M. DELANEY, AND M. TILTON. 2007. Carbon storage and Oregon's land-use planning program. *Journal of Forestry* 105(5):167–172.

CAYAN, D., A.L. LUERS, M. HANEMANN, G. FRANCO, AND B. CROES. 2005. *Possible scenarios of climate change in California and recommendations.* CEC-500-2005-186-SD. California Climate Change Center.

CESS, R.D., G.L. POTTER, S.J. GHAN, AND W.L. GATES. 1985. The climatic effects of large injections of atmospheric smoke and dust: A study of climate feedback mechanisms with one- and three-dimensional climate models. *Journal of Geophysical Research* 90:12937–12950.

CHAMBERS, J.Q., J.I. FISHER, H. ZENG, E.L. CHAPMAN, D.B. BAKER, AND G.C. HURTT. 2007. Hurricane Katrina's carbon footprint on U.S. Gulf Coast forests. *Science* 16:1107.

CLARK, J.S. 1998. Why trees migrate so fast: Confronting theory with dispersal biology and the paleorecord. *The American Naturalist* 152(2):204–224.

CLEAN AIR–COOL PLANET. 2006. *A consumer's guide to retail carbon offset providers*. Available online at www.cleanair-coolplanet.org/ConsumersGuidetoCarbonOffsets. pdf; accessed October 12, 2007.

CLIMATE LEADERSHIP INITIATIVE. 2007. *Economic impacts of climate change on forest resources in Oregon.* Institute for a Sustainable Environment, University of Oregon.

CLUTTER, M., B. MENDELL, D. NEWMAN, D. WEAR, AND J. GREIS. 2005. *Strategic factors driving timberland ownership changes in the South.* USDA Forest Service, Southern Research Station, Research Triangle Park, NC.

COLLEGE OF FOREST RESOURCES (CFR). 2007. *Study 1: Timber supply and forest structure in the future of Washington's forests and forestry industries.* Final report by the College of Forest Resources, University of Washington, Seattle.

CUBBAGE, F., D. GREENE, AND J. LYON. 1989. Tree size and species, stand volume and tract size: Effects on southern harvesting costs. *Southern Journal of Applied Forestry* 13(3):145–152.

CUSHING, T., S. BULLARD, J. GREENE, AND T. BEAUVIAS. 1998. The effects of the federal estate tax on nonindustrial private landowners. *Proceedings of the Society of American Foresters 1998 National Convention*, Traverse City, Michigan, September 19–23, 1998.

DALE, V.H., L.A. JOYCE, S. MCNULTY, R.P. NEILSON, M.P. AYRES, M.D. FLANNIGAN, P.J. HANSON, L.C. IRLAND, A.E. LUGO, C.J. PETERSON, D. SIMBERLOFF, F.J. SWANSON, B.J. STOCKS, AND M. WOTTON. 2001. Climate change and forest disturbances. *BioScience* 51(9):723–734.

DALY, C., J.W. SMITH, J.I. SMITH, AND R.B.MCKANE. 2007. High-resolution spatial scale modeling of daily weather elements for a catchment in the Oregon Cascade mountains, United States. *Journal of Applied Meteorology and Climatology* 47:1565–1586.

DANIELS, T. 1991. The purchase of development rights: Preserving agricultural land and open space. *Journal of the American Planning Association* 57(4):421–431.

DANIELS, T., AND M. LAPPING. 2005. Land preservation: An essential ingredient in smart growth. *Journal of Planning Literature* 19(3): 316–329.

DARLEY, E.E. 1979. *Emission factors from hydrocarbon characterization of agricultural waste burning.* CAL/ARB Project A7–068-30. University of California, Riverside.

DEWALLE, D.R., A.R. BUDA, AND A. FISHER. 2003. Extreme weather and forest management in the mid-Atlantic region of the United States. *National Journal of Applied Forestry* 20(2):61–70.

DIXON, R.K., AND O.N. KRANKINA. 1993. Forest fires in Russia: Carbon dioxide emissions to the atmosphere. *Canadian Journal of Forest Research* 23:700–705.

DIXON, R.K., A.M. SOLOMON, S. BROWN, R.A. HOUGHTON, M.C. TREXIER, AND J. WISNIEWSKI. 1994. Carbon pools and flux of global forest ecosystems. *Science* 263(5144):185–190.

EASTERLING, W.E., P.K. AGGARWAL, P. BATIMA, K.M. BRANDER, L. ERDA, S.M. HOWDEN, A. KIRILENKO, J. MORTON, J.-F. SOUSSANA, J. SCHMIDHUBER, AND F.N. TUBIELLO. 2007. Food, fibre and forest products. In *Climate change 2007: Impacts, adaptation and vulnerability. Contribution of Working Group II to the Fourth Assessment Report of the Intergovernmental Panel on Climate Change*, Parry, M.L., O.F. Canziani, J.P. Palutikof, P.J. van der Linden, and C.E. Hanson (eds.). Cambridge University Press, Cambridge, UK.

EATOUGH-JONES, M., T.D. PAINE, M.E. FENN, AND M.A. POTH. 2004. Influence of ozone and nitrogen deposition on bark beetle activity under drought conditions. *Forest Ecology and Management* 200:67–76.

ENERGY INFORMATION ADMINISTRATION (EIA). 2006. Renewable energy consumption and electricity preliminary statistics. Available online at www.eia.doe.gov/cneaf/solar.renewables/page/prelim_trends/rea_prereport.html; accessed October 14, 2007.

ENGLISH, B.C., D.G. DE LA TORRE UGARTE, K. JENSEN, C. HELLWINCKEL, J. MENARD, B. WISON, R. ROBERTS, AND M.WALSH. 2006. *25% renewable energy for the United States by 2025: Agricultural and economic impacts.* University of Tennessee, Knoxville.

EUROPEAN COMMISSION. 2005. *EU action against climate change: EU emissions trading—An open scheme promoting innovations.* Brussels, Belgium.

FIELD, C.B., L.D. MORTSCH, M. BRKLACICH, D.L. FORBES, P. KOVACS, J.A. PATZ, S.W. RUNNING, AND M.J. SCOTT. 2007. North America. In *Climate change 2007: Impacts, adaptation and vulnerability. Contribution of Working Group II to the Fourth Assessment Report of the Intergovernmental Panel on Climate Change,* Parry, M.L., O.F. Canziani, J.P. Palutikof, P.J. van der Linden, and C.E. Hanson (eds.). Cambridge University Press, Cambridge, UK.

FINNEY, M.A. 2000. Design of regular landscape fuel treatment patterns for modifying fire growth and behavior. *Forest Science Journal* 47(2):219–228.

FLANNIGAN, M.D., Y. BERGERON, O. ENGELMARK, AND B.M. WOTTON. 1998. Future wildfire in circumboreal forests in relation to global warming. *Journal of Vegetation Science* 9(4):469–476.

FLEMING, R.A., J.N. CANDAU, AND R.S. MCALPINE. 2002. Landscape-scale analysis of interactions between insect defoliation and forest fire in central Canada. *Climatic Change* 55:251–272.

FRANZLUEBBERS, A.J. 2005. Soil organic carbon sequestration and agricultural greenhouse gas emissions in the southeastern USA. *Soil and Tillage Research* 83:120–147.

GAN, J. 2004. Risk and damage of pine beetle outbreaks under global climate change. *Forest Ecology and Management* 191:61–71.

GAYER, P.D., H.L. HANEY JR., AND C.A. HICKMAN. 1987. *Effect of current-use valuation on forestry investment returns in selected Virginia counties.* Res. Note SO-332. USDA Forest Service, Southern Forest Experiment Station, New Orleans.

GERRARD, M. 2007. *Global climate change and U.S. law.* American Bar Association, Washington, DC.

GLOBAL FIRE MONITORING CENTER (GFMC). 2003. Russian Federation Fire 2002 Special, the wildland fire season 2002 in the Russian Federation—An assessment by the Global Fire Monitoring Center. *International Fire News* 28 (January–June 2003). Available online at www.fire.uni-freiburg.de/iffn/country/rus/ IFFN%20Russia%202002%20Fire%20 Report.pdf; accessed October 16, 2007.

GOLDAMMER, J.G. (ED.). 1990. *Fire in the tropical biota: Ecosystem processes and global challenges.* Ecological Studies 84, Springer-Verlag, Berlin.

GORTE, R.W. 2007. *Carbon sequestration in forests.* CRS Report for Congress RL31432. Congressional Research Service.

GOVERNMENT ACCOUNTABILITY OFFICE (GAO). 2007. *Climate change: Agencies should develop guidance for addressing the effects and federal land and water resources.* GAO-07-863. Washington, DC.

HALPIN, P.N. 1997. Global climate change and natural-area protection: Management responses and research directions. *Ecological Applications* 7(3):828–843.

HAMILTON, J.G., E.H. DELUCIA, K. GEORGE, S.L. NAIDU, A.C. FINZI, AND W.H. SCHLESINGER. 2002. Forest carbon balance under elevated CO_2. *Oecologia* 131:250–260.

HAMILTON, K., R. BAYON, G. TURNER, AND D. HIGGINS. 2007. *State of the voluntary carbon market 2007—Picking up steam.* Ecosystem Marketplace, Washington, DC.

HAMILTON, K., M. SJARDIN, T. MARCELLO, AND G. XU. 2008. *Forging a frontier: State of the voluntary carbon market 2008.* Ecosystem Marketplace and New Carbon Finance, Washington, DC.

HAMMERSCHLAG, R. 2006. Ethanol's energy return on investment: A survey of the literature 1990–present. *Environmental Science and Technology* 40(6):1744–1750.

HANDA, I.T., C. KORNER, AND S. HATTENSCHWILER. 2005. A test of the treeline carbon limitation hypothesis by in situ CO_2 enrichment and defoliation. *Ecology* 86(5):1288–1300.

HANLEY, N., J.F. SHOGREN, AND B.WHITE. 1997. *Environmental economics in theory and practice.* Oxford University Press, New York.

HARMON. M.E. 2006. Atmospheric Carbon Dioxide. In *Forests, carbon and climate change: A synthesis of science findings.* Oregon Resources Institute, Portland, OR.

HARMON, M.E., W.K. FERRELL, AND J.F. FRANKLIN. 1990. Effects on carbon storage of conversion of old-growth forests to young forests. *Science* 247(4943):699–702.

HASSE, S. 2007. *Wood to energy.* Presentation to Southwest Sustainable Forest Partnership Small Wood Entrepreneurial Conference, Ruidoso, NM, November.

HAYNES, R.W. 2003. *An analysis of the timber situation in the United States: 1952 to 2050.* PNW-GTR-560. USDA Forest Service, Pacific Northwest Research Station, Portland, OR.

HAYNES, R.W., D.M. ADAMS, R.J. ALIG, P.J. INCE, J.R.MILLS, AND X. ZHOU. 2007. *The 2005 RPA timber assessment update.* PNWGTR-699. USDA Forest Service, Pacific Northwest Research Station, Portland, OR.

HEATH, L.S., AND R.A. BIRDSEY. 1993. Carbon trends of productive temperate forests of the coterminous United States. *Water, Air, and Soil Pollution* 70:279–293.

HEATH, L.S., AND J.E. SMITH. 2000. Soil carbon accounting and assumptions for forestry and forestry-related land use change. In *The impact of climate change on America's forests,* Joyce, L.A., and R.A. Birdsey (eds.).RMRSGTR-59. USDA Forest Service, Rocky Mountain Research Station, Fort Collins, CO.

HEATH, L.S., R.A. BIRDSEY, C. ROW, AND A.J. PLANTINGA. 1996. Carbon pools and flux in U.S. forest products. In *Forest ecosystems, forest management, and the global carbon cycle,* Apps, M.J., and D.T. Price (eds.). NATO ASI Series I: Global Environmental Changes, Volume 40, Springer-Verlag.

HEATH, L.S., J.M. KIMBLE, R.A. BIRDSEY, AND R. LAL. 2003. The potential of U.S. forest soils to sequester carbon. In *The potential of U.S. forest soils to sequester carbon and mitigate the greenhouse effect,* Kimble, J.M., L.S. Heath, R.A. Birdsey, and R. Lal (eds.). CRC Press, Baton Rouge, FL.

HEATH, L.S., J.A. SMITH, AND R.A. BIRDSEY. 2003. Carbon trends in U.S. forestlands: A context for the role of forest soils in forest carbon sequestration. In *The potential of U.S. forest soils to sequester carbon and mitigate the greenhouse effect,* Kimble, J.M., L.S. Heath, R.A. Birdsey, and R. Lal (eds.), CRC Press, Baton Rouge, FL.

HELMS, J.A. 1965. Diurnal and seasonal patterns of net assimilation in Douglas-fir (*Pseudotsuga menziesii*) as influenced by environment. *Ecology* 46(5):698–708.

HELMS, J.A. 1996. Management strategies for sequestering and storing carbon in natural, managed, and highly disturbed coniferous ecosystems. Proceedings of Conference on Forests and Global Change, Washington, DC, June 11–12. In *Forests and global change. Vol. 2. Forest management opportunities for mitigating carbon emissions,* Sampson, R.N., and D. Hair (eds.). American Forests, Washington, DC.

HOBBS, R.J., S. ARICO, J. ARONSON, J.S. BARON, P. BRIDGEWATER, A.A. CRAMER, P.R. EPSTEIN, J.J. EWEL, C.A. KLINK, A.E. LUGO, D. NORTON, D. OJIMA, D.M. RICHARDSON, E.W. SANDERSON, R. VALLADARES, M. VILA, R. ZAMORA, AND M. ZOBEL. 2006. Novel ecosystems: Theoretical and management aspects of the new ecological world order. *Global Ecology and Biogeography* 15(1):1–7.

HOLLAND, E.A., S. BROWN, C.S. POTTER, S.A. KLOOSTER, S. FAN,M. GLOOR, J.MAHLMAN, S. PACALA, J. SARMIENTO, T. TAKAHASHI, AND P. TANS. 1999. North American carbon sink. *Science* 283(5409):1815a.

HOOVER, C., AND S. STOUT. 2007. The carbon consequences of thinning techniques: Stand structure makes a difference. *Journal of Forestry* 105(5):266–270.

INGERSON, A. 2007. *U.S. forest carbon and climate change: Controversies and win-win policy approaches.* Wilderness Society, Washington, DC.

INTERGOVERNMENTAL PANEL ON CLIMATE CHANGE (IPCC). 1992. *Climate change 1992: The Supplementary Report to the IPCC Scientific Assessment,* Houghton, J.T. (ed.). Cambridge University Press, New York.

INTERGOVERNMENTAL PANEL ON CLIMATE CHANGE (IPCC). 2006. Guidelines for national greenhouse gas inventories, Chapter 12, Harvested wood products. Available online at www.ipcc-nggip.iges.or.jp/public/2006gl/ index.htm; accessed November 13, 2007.

INTERGOVERNMENTAL PANEL ON CLIMATE CHANGE (IPCC). 2007. *Climate change 2007: The physical science basis. Contribution of Working Group I to the Fourth Assessment Report of the Intergovernmental Panel on Climate Change,* Solomon, S., D. Qin, M. Manning, Z. Chen, M. Marquis, K.B. Averyt, M. Tignor, and H.L. Miller (eds.). Cambridge University Press, Cambridge, UK.

INTERNATIONAL EMISSIONS TRADING ASSOCIATION (IETA). 2007. *IETA's guidance note through the CDM project approval process, v. 3.0.* Available online at www.ieta.org/ieta/www/pages/getfile.php?docID_2370; accessed October 15, 2007.

IRLAND, L.C., D. ADAMS, R. ALIG, C.J. BETZ, C.-C. CHEN,M. HUTCHINS, B.A.MCCARL, K. SKOG, AND B.L. SOHNGEN. 2001. Assessing socioeconomic impacts of climate change on U.S. forests, wood-product markets, and forest recreation. *BioScience* 51(9):753–764.

IVERSON, L.R., AND A.M. PRASAD. 2002. Potential redistribution of tree species habitat under five climate change scenarios in the eastern U.S. *Forest Ecology and Management* 155:205–222.

JOYCE, L.A., J.R. MILLS, L.S. HEATH, A.D. MCGUIRE, R.W. HAYNES, AND R.A. BIRDSEY. 1995. Forest sector impacts from changes in forest productivity under climate change. *Journal of Biogeography* 22(4/5):703–713.

KAUFMAN, Y.J., I. KOREN, L.A. REMER, D. ROSENFELD, AND Y. RUDICH. 2005. The effect of smoke, dust, and pollution aerosol on shallow cloud development over the Atlantic Ocean. *Proceedings of the National Academy of Sciences of the United States of America* 102(32): 11207–11212.

KAUFMANN, M.R., A. SHLISKY, AND P. MARCHAND. 2005. *Good fire, bad fire: How to think about forest land management and ecological processes.* USDA Forest Service, Rocky Mountain Research Station, Fort Collins, CO.

KELTY, M.J. 2006. The role of species mixtures in plantation forestry. *Forest Ecology and Management* 233:195-204.

KIMBLE, J.M., R.A. BIRDSEY, R. LAL, AND L.S. HEATH. 2003. Introduction and general description of U.S. forests. In *The potential of U.S. forest soils to sequester carbon and mitigate the greenhouse effect,* Kimble, J.M., L.S. Heath, R.A. Birdsey, and R. Lal (eds.). CRC Press, Baton Rouge, FL.

KING, A.W., L. DILLING, G.P. ZIMMERMAN, D.M. FAIRMAN, R.A. HOUGHTON, G. MARLAND, A.Z. ROSE, AND T.J. WILBANKS (EDS.). 2007. *North American carbon budget and implications for the global carbon budget.* US Climate Change Science Program. Synthesis and Assessment Product 2.2. National Oceanic and Atmospheric Administration, National Climatic Data Center, Asheville, NC.

KIRSCHBAUM, M.U.F. 2000. Forest growth and species distribution in a changing climate. *Tree Physiology* 20:309–322.

KORNER, C., R. ASSHOFF, O. BIGNUCOLO, S. HATTENSCHWILER, S.G. KEEL, S. PELAEZRIEDL, S. PEPIN, R.T.W. SIEGWOLF, AND G. ZOTZ. 2005. Carbon flux and growth in mature deciduous forest trees exposed to elevated CO_2. *Science* 309:1360–1362.

KRAMER, P.J. 1981. Carbon dioxide concentration, photosynthesis, and dry matter production. *BioScience* 31(1):29–33.

KRANKINA, O.M., AND M. HARMON. 2006. Forest management strategies for carbon storage. In *Forests, carbon and climate change: A synthesis of science findings.* Oregon Forest Resources Institute, Portland, OR.

LAL, L. 2005. Forest soils and carbon sequestration. *Forest Ecology and Management* 220:242–258.

LARSON, E.D., S. CONSONNI, R.E. KATOFSKY, K. LISA, AND J.W.J. FREDERICK. 2006. *A cost-benefit assessment of gasification-based biorefining in the kraft pulp and paper industry. Volume 1.* Princeton University, Princeton, NJ.

LEGGETT, J.A. 2007. *Climate change: Science and policy implications.* Congressional Research Service, Washington, DC.

LINCOLN, D.E., E.D. FAJER, AND R.H. JOHNSON. 1993. Plant-insect herbivore interactions in elevated CO_2 environments. *TREE* 8(2):64–68.

LIPPKE, B. 2007. *Climate change, carbon and the forest sector.* Presentation to American Forest Resource Council Annual Meeting, Skamania, WA,April 11. Available online at www.corrim. org/ppt/2007/lippke_afrc/index.htm; accessed November 13, 2007.

LIPPKE, B., AND L. EDMONDS. 2006. Environmental performance improvements in residential construction: The impact of products, biofuels, and processes. *Forest Products Journal* 56(10):58–63.

LIPPKE, B., J. COMNICK, AND C.L.MASON. 2006. Alternative landscape fuel removal scenarios: Impacts of thinning treatment intensity and implementation schedules on fire hazard reduction effectiveness, carbon storage, and economics. RTI/CORRIM Joint Working Paper No. 6. Seattle.

LIPPKE, B., J. WILSON, J. PEREZ-GARCIA, J. BOWYER, AND J. MEIL. 2004. CORRIM: Life-cycle environmental performance of renewable building materials. *Forest Products Journal* 54(6):8–19.

LOGAN, J.A., AND J.A. POWELL. 2001. Ghost forests, global warming, and the mountain pine beetle. *American Entomologist* 47(3):160–173.
LOGAN, J.A., J. RE ´ GNIE`RE, AND J.A. POWELL. 2003. Assessing the impacts of global warming on forest pest dynamics. *Frontiers in Ecology and Environment* 1(3):130–137.
LOOMIS, J., P.WOHLGEMUTH, A. GONZALEZ-CABAN, AND D. ENGLISH. 2003. Economic benefits of reducing fire-related sediment in southwestern fire prone ecosystems. *Water Resources Research* 39(9):1260.
LYND, L.R., AND M. WANG. 2004. A productnonspecific framework for evaluating the potential of biomass-based products to displace fossil fuels. *Journal of Industrial Ecology* 7(3–4):17–32.
MALHI, Y., AND J. GRACE. 2000. Tropical forests and atmospheric carbon dioxide. *Trends in Ecology and Evolution* 15(8):332–337.
MALMSHEIMER, R.W., P. HEFFERNAN, S. BRINK, D. CRANDALL, F. DENEKE, C. GALIK, E. GEE, J.A. HELMS, N. MCCLURE, M. MORTIMER, S. RUDDELL, M. SMITH, and J. STEWART. Forest Management Solutions for Mitigating Climate Change in the United States. *Journal of Forestry* 106(3):115-173.
MARLAND, G., AND B. SCHLAMADINGER. 1999. Forests for carbon sequestration or fossil fuel substitution? A sensitivity analysis. *Biomass and Bioenergy* 13(6):389–397.
MASON, L., B. LIPPKE, K. ZOBRIST, T. BLOXTON, K. CEDER, J. COMNICK, J. MCCARTER, AND H. ROGERS. 2006. Investment in fuel removals to avoid fires result in substantial benefits. *Journal of Forestry* 104(1):27–31.
MCHALE, M.R., E.G. MCPHERSON, AND I.C. BURKE. 2007. The potential of urban tree plantings to be cost effective in carbon credit markets. *Urban Forestry and Urban Greening* 6(1):49–60.
MCKITRICK, R., J. D'ALEO, M. KHANDEKAR, W. KININMONTH, C. ESSEX, W. KARLE´N, O. KA¨RNER, I. CLARK, T. MURTY, AND J.J. O'BRIEN. 2007. Independent summary for policymakers: IPCC Fourth Assessment Report. Fraser Institute, Vancouver, BC.
MCLACHLAN, J.S., J.J. HELLMANN, AND M.W. SCHWARTZ. 2007. A framework for debate of assisted migration in an era of climate change. *Conservation Biology* 21(2):297–302.
MCMAHON, J.P. 2007. *Testimony by John P. Mc- Mahon, President, Society of American Foresters*, House Committee on Appropriations, Subcommittee on Interior and Related Agencies, concerning Fiscal Year 2008 Interior, Environment and Related Agencies Budget. April 19.
MILLAR, C.I., N.L. STEPHENSON, AND S.L. STEPHENS. 2007. Climate change and forests of the future: Managing in the face of uncertainty. *Ecological Applications* 17(8):2145–2151. Available online at www.fs.fed.us/psw/publications/millar/psw_2007_millar029.pdf; accessed February 7, 2008.
MORRIS, G. 2007. *Greenhouse gas white paper—Preliminary results*. Green Power Institute, Berkeley, CA.
MORTIMER, M.J., J.J. RICHARDSON JR., J.S.HUFF, AND H.L. HANEY JR. 2007. A survey of forestland conservation easements in the United States: Implications for the management and stewardship of working private forests. *Small- Scale Forest Economics, Management, and Policy* 6(1):35–47.
MORTIMER, M.J., L. STULL, S.P. PRISLEY, AND D. SLACK. 2006. Forest-related ordinances in Virginia: A case study in regulatory de-evolution. *Southern Journal of Applied Forestry* 30(4):196–205.
MURRAY, B.C., B.A. MCCARL, AND H.C. LEE. 2004. Estimating leakage for forest carbon sequestration programs. *Land Economics* 80(1):109–124.
NABUURS, G.J., O.MASERA, K. ANDRASKO, P. BENITEZ- PONCE, R. BOER, M. DUTSCHKE, E. ELSIDDIG, J. FORD-ROBERTSON, P. FRUMHOFF, T. KARJALAINEN, O. KRANKINA, W.A. KURZ, M. MATSUMOTO, W. OYHANTCABAL, N.H. RAVINDRANATH, M.J.S. SANCHEZ, AND X. ZHANG. 2007. Ch. 9. Forestry. In *Climate change 2007: Mitigation. Contribution of Working Group lll to the Fourth Assessment Report of the Intergovernmental Panel on Climate Change*, Metz, B., O.R. Davidson, P.R. Bosch, R. Dave and L.A. Meyer (eds.). Cambridge University Press, Cambridge, UK.
NABUURS, G.J., A. PUSSINEN, T. KARJALAINEN,M. ERHARD, AND K. KRAMER. 2002. Stemwood volume increment changes in European forests due to climate change—A simulation study with the EFISCEN model. *Global Change Biology* 8:304–316.
NATIONAL INTERAGENCY FIRE CENTER (NIFC). 2007. *Wildland fire statistics*. Available online at www.nifc.gov/fire_info/fire_stats.htm; accessed November 9, 2007
NATIONAL RENEWABLE ENERGY LABORATORY (NREL). 2006. *From biomass to biofuels: NREL leads the way*. NREL/BR-510-39436. Available online at www.nrel.gov/biomass/pdfs/39436.pdf; accessed February 6, 2008

NATIONAL RESEARCH COUNCIL (NRC). 2006. *Surface temperature reconstructions for the last 2,000 years*. National Academies Press, Washington, DC.

NATURAL RESOURCES CONSERVATION SERVICE (NRCS). 2001. Natural resources inventory. US Department of Agriculture, Washington, DC.

NATURAL RESOURCES CONSERVATION SERVICE (NRCS). 2007. *Natural resources inventory. 2003 Annual NRI—Land use*. US Department of Agriculture, Washington, DC.

NEILSON, R.P. 2007. *Climate change, uncertainty and forecasts of global to landscape ecosystem dynamics (percent change in biomass burned)*. Presentation to Wildland Fire Leadership Council, Red Lodge, Montana, June 20.

NEUENSCHWANDER, L.F., J.P. MENAKIS, M. MILLER, R.N. SAMPSON, C. HARDY, B. AVERILL, AND R. MASK. 2000. Indexing Colorado watersheds to risk of wildfire. *Journal of Sustainable Forestry* 11(1–2):35–55.

NITSCHKE, C.R., AND J.L. INNES. 2006. Interactions between fire, climate change and forest biodiversity. *Perspectives in Agriculture, Veterinary Science, Nutrition and Natural Resources* 1(60):9.

NORBY, R.J., E.H. DELUCIA, B. GIELEN, C. CALFAPIETRA, C.P. GIARDINA, J.S. KING, J. LEDFORD, H.R. MCCARTHY, D.J.P. MOORE, R. CEULEMANS, P.D. ANGELIS, A.C. FINZI, D.F. KARNOSKY, M.E. KUBISKE, M. LUKAC, K.S. PREGITZER, G.E. SCARASCIA-MUGNOZZA, W.S. SCHLESINGER, AND R. OREN. 2005. Forest response to elevated CO_2 is conserved across a broad range of productivity. *Proceedings of the National Academy of Sciences* 102(50):18052–18056.

NORBY, R.J., P.J. HANSON, E.G. O'NEILL, T.J. TSCHAPLINSKI, J.F. WELTZIN, R.A. HANSEN, W. CHENG, S.D. WULLSCHLEGER, C.A. GUNDERSON, N.T. EDWARDS, AND D.W. JOHNSON. 2002. Net primary productivity of a CO_2-enriched deciduous forest and the implications for carbon storage. *Ecological Applications* 12(5):1261–1266.

NORBY, R.J., J. LEDFORD, C.D. REILLY, N.E. MILLER, E.G. O'NEILL, AND W.H. SCHLESINGER. 2004. Fine-root production dominates response of a deciduous forest to atmospheric CO_2 enrichment. *Proceedings of the National Academy of Sciences* 101(26):9689–9693.

NORBY, R.J., S.D. WULLSCHLEGER, C.A. GUNDERSON, D.W. JOHNSON, AND R. CEULEMANS. 1999. Tree responses to rising CO_2 in field experiments: Implications for the future forest. *Plant, Cell and Environment* 22:683–714.

NOSS, R.F. 2001. Beyond Kyoto: Forest management in a time of rapid climate change. *Conservation Biology* 15(3):578–590.

NOWAK, D.J., AND D.E. CRANE. 2002. Carbon storage and sequestration by urban trees in the USA. *Environmental Pollution* 116:381–389.

ONEIL, E. 2007. Changes in landscape level carbon pools over time: Appendix 2. Final report review draft. In *CORRIM: Phase II final report module A: Forest resources Inland West*, Johnson, L., B. Lippke, and E. Oneil (eds.). Consortium for Research on Renewable Industrial Materials, Seattle.

ONEIL, E., B. LIPPKE, AND L.MASON. 2007. Eastside climate change, forest health, fire, and carbon accounting. Discussion paper 8. In *The future of Washington's forests and forestry industries. Final report.* Prepared by University of Washington College of Forest Resources for Washington State Department of Natural Resources.

ONIANWA, O., AND G.WHEELOCK. 1999. Assessing the retention potential of Conservation Reserve Program practices in Alabama. *Southern Journal of Applied Forestry* 23(2):83–87.

OPDAM, P., AND D. WASCHER. 2004. Climate change meets habitat fragmentation linking landscape and biogeographical scale levels in research and conservation. *Biological Conservation* 117:285–297.

OREGON FOREST RESOURCES INSTITUTE (OFRI). 2006. *Forests, carbon and climate change: A synthesis of science findings*. Available online at http://www.oregonforests.org/media/pdf/CarbonRptFinal.pdf; accessed February 3, 2008

OREN, R., D.S. ELLSWORTH, K.H. JOHNSEN, N. PHILLIPSK, B.E. EWERS, C. MAIER, K.V.R. SCHAEFER, H. MCCARTHY, G. HENDREY, S.G. MCNULTY, AND G.G. KATUL. 2001. Soil fertility limits carbon sequestration by forest ecosystems in a CO_2-enriched atmosphere. *Nature* 411:469–472.

PACHECO,M. 2005. DOE-NREL presentation to 25x'25 Renewable Energy Forestry Work Group, Tucson, AZ, January.

PARMESAN, C., AND G. YOHE. 2003. A globally coherent fingerprint of climate change impacts across natural systems. *Nature* 421(6918):37–42.

PARTRIDGE, C., AND B. MACGREGOR. 2007. *The future of Washington's forests.* Washington State Department of Natural Resources, Olympia.

PEARCE, F. 2005. Clearing smoke may trigger global warming rise. *New Scientist Environment.* Available online at environment.newscientist.com/article/dn7607.html; accessed October 11, 2007.

PEREZ-GARCIA, J., B. LIPPKE, J. COMNICK, AND C. MANRIQUEZ. 2005. An assessment of carbon pools, storage, and wood products market substitution using life-cycle analysis results. *Wood and Fiber Science* 37:140–148.

PERLACK, R.D., L.L.WRIGHT, A.F. TURHOLLOW, R.L.GRAHAM, B.J. STOKES, AND D.C. ERBACH. 2005. *Biomass as feedstock for a bioenergy and bioproducts industry: The technical feasibility of a billion-ton annual supply.* US Department of Energy, Oak Ridge National Laboratory, Oak Ridge, TN.

PERMAN, R., Y. MA, AND J. MCGILVARY. 1996. *Natural resource and environmental economics.* Longman Publishing, New York.

PERSCHEL, R.T., A.M. EVANS, AND M.J. SUMMERS. 2007. *Climate change, carbon, and the forests of the Northeast.* Forest Guild. Santa Fe, NM. Available online at www.forestguild.org/publications/2007/ForestGuild_climate_ carbon_forests.pdf; accessed February 6, 2008

PETERSON, D.L., M.C. JOHNSON, J.K. AGEE, T.B. JAIN, D. MCKENZIE, AND E.D. REINHARDT. 2005. *Forest structure and fire hazard in dry forests of the western United States.* PNW-GTR- 628. USDA Forest Service, Pacific Northwest Research Station.

PIMENTEL, D., AND T. PATZEK. 2005. Ethanol production using corn switchgrass, and wood; Biodiesel production using soybean and sunflower. *Natural Resources Research* 14(1):65–76.

POINT CARBON NEWS. 2007. *Carbon market North America.* Washington, DC. 2(22):7.

POUNDS, J.A., M.P.L. FOGDEN, AND J.H. CAMPBELL. 1999. Biological response to climate change on a tropical mountain. *Nature* 398:611–615.

PREGITZER, K.S., AND E.S. EUSKIRCHEN. 2004. Carbon cycling and storage in world forests: Biome patterns related to forest age. *Global Change Biology* 10:2052–2077.

PRICE, T.D., M.J. APPS, AND W.A. KURZ. 1998. Past and possible future carbon dynamics of Canada's boreal forest ecosystems. In *Carbon dioxide in forestry and wood industry,* Kohlmaier, G.K., M. Weber, and R.A. Houghton (eds.). Springer-Verlag, Berlin.

PRISLEY, S.P., D.R. DAVERSA, AND M.J. MORTIMER. 2006. Estimation of forest area affected by local ordinances: A Virginia case study. *Southern Journal of Applied Forestry* 30(4):188–195.

PUETTMANN, M., AND J. WILSON. 2005. Life-cycle analysis of wood products: Cradle to gate LCI of residential wood building materials. *Wood and Fiber Science* 37:18–29.

REAY, D.S., S. RADAJEWSKI, J.C. MURRELL, N. MCNAMARA, AND D.B. NEDWELL. 2001. Effects of land-use on the activity and diversity of methane oxidizing bacteria in forest soils. *Soil Biology and Biochemistry* 33(12–13):1613–1623.

REHFELDT, G.E., N.L. CROOKSTON, M.V. WARWELL, AND J.S. EVANS. 2006. Empirical analyses of plant-climate relationships for the western United States. *International Journal of Plant Sciences* 167(6):1123–1150. Available online at www.fs.fed.us/rm/pubs_other/rmrs_2006_rehfeldt_g001.pdf; accessed February 6, 2008.

RICHARDS, K.R., R. MOULTON, AND R.A. BIRDSEY. 1993. Costs of creating carbon sinks in the U.S. *Energy Conservation and Management* 34(9-11):905–912.

ROJ, A. 2005. *Automotive fuels from biomass: What is the best road forward.* Presentation to First International Biorefinery Workshop, Washington, DC, July.

RUDDELL, S., R.N. SAMPSON, M. SMITH, R.A. GIFFEN, J. CATHCART, J.M. HAGAN, D.L. SOSLAND, J. HEISSENBUTTEL, J.F. GODBEE, S.M. LOVETT, J.A. HELMS, W.C. PRICE, AND R.S. SIMPSON. 2007. The role for sustainably managed forests in climate change mitigation. *Journal of Forestry* 105(6):314–319.

RUDDELL, S., M.J.WALSH, AND M. KANAKASABAI. 2006. *Forest carbon trading and marketing in the United States.* Report to the North Carolina Division of the Society of American Foresters. Available online at www.foreconecomarketsolutionsllc.com.

RYAN, K.C. 2000. Global change and wildland fire. In *Wildland fire in ecosystems: Effects of fire on flora,* Brown, J.K., and J.K. Smith (eds.). GTR-42, vol. 2. USDA Forest Service, Rocky Mountain Research Station, Fort Collins, CO.

SACKS, W.J., D.S. SCIMEL, AND R.K. MONSON. 2007. Coupling between carbon cycling and climate in a high-elevation subalpine forest: A model-data fusion analysis. *Oecologia* 151:54–68.

SAMPSON, R.N., AND R.J. SCHOLES. 2000. Additional human-induced activities—Article 3.4. In *Land use, land use change and forestry,* Watson, R.T., I.R. Noble, B. Bolin, N.H. Ravindranath, D.J. Verardo, and D.J. Dokken (eds.). Intergovernmental Panel on Climate Change special report. Available online at www.grida.no/climate/ipcc/land_use; accessed November 17, 2007.

SAMPSON, R.N. 2004. *Experience with fire and fire management in U.S. forests and the potential to increase carbon stocks on forest land by changing practices.* Available online at www.sampsongroup.com/Papers/Fire%20Management,%20Biomass,%20and%20Carbon.pdf; accessed October 16, 2007.

SAMPSON, R.N., AND M.H. KAMP. 2007. Terrestrial carbon sequestration in the Northeast: Quantities and costs. Part 2—Recent trends in sinks and sources of carbon. Available online at conserveonline.org/workspaces/necarbonproject/The%20Report/Part%202%20–%20Recent%20Trends.pdf/view; accessed November 29, 2007.

SAMPSON, R.N., S. RUDDELL, AND M. SMITH (EDS.). 2007. *Managed forests in climate change policy: Program design elements.* Available online at www.safnet.org/policyandpress/documents/managedforests_12–14-07.pdf; accessed February 11, 2008.

SATHRE, R. 2007. *Life-cycle energy and carbon implications of wood-based products and construction.* Mid-Sweden University doctoral thesis 34. Ostersund, Sweden.

SAUER, A. 2002. *The value of conservation easements: The importance of protecting nature and open space.* Discussion paper, April 9. World Resources Institute, Washington, DC.

SAXE, H., D.S. ELLSWORTH, AND J.HEATH. 1998. Tree and forest functioning in an enriched CO_2 atmosphere. *New Phytologist* 139(3): 395–436.

SCHLAMADINGER, B., AND G. MARLAND. 1996. The role of forest and bioenergy strategies in the global carbon cycle. *Biomass and Bioenergy* 10(5–6):275–300.

SCHLYTER, P., I. STJERNQUIST, L. BA ¨RRING, A.M. JO ¨ NSSON, AND C. NILSSON. 2006. Assessment of the impacts of climate change and weather extremes on boreal forests in northern Europe, focusing on Norway spruce. *Climate Research* 31:75–84.

SCHRAG, A.M., A.G. BUNN, AND L.J. GRAUMLICH. 2007. Influence of bioclimatic variables on tree line conifer distribution in the Greater Yellowstone Ecosystem: Implications for species of conservation concern. *Journal of Biogeography.* Online early articles doi: 10.1111/j. 1365-2699.2007.01815.x. Available online at www.blackwell-synergy.com/doi/abs/10.1111/j.1365-2699.2007.01815.x; accessed February 6, 2008.

SCHULZE, E., C. WIRTH, AND M. HEIMANN. 2000. Managing forests after Kyoto. *Science* 289(5487):2058–2059.

SCHWARTZ, M.W. 1993. Modeling effects of habitat fragmentation on the ability of trees to respond to climatic warming. *Biodiversity and Conservation* 2:51–61.

SCHWARTZ, M.W., L.R. IVERSON, AND A.M. PRASAD. 2001. Predicting the potential future distribution of four tree species in Ohio using current habitat availability and climatic forcing. *Ecosystems* 4(6):568–581.

SHEEHAN, J., A. ADEN, K. PAUSTAIN, K. KILLIAN, J. BRENNER, M. WALSH, AND R. NELSON. 2004. Energy and environmental aspects of using corn stover for fuel ethanol. *Journal of Industrial Ecology* 7(3–4):17–32.

SHUGART, H., R. SEDJO, AND B. SOHNGEN. 2003. *Forests and global climate change: Potential impacts on U.S. forest resources.* Pew Center on Global Climate Change, Washington, DC.

SINGER, S.F. (ED.). 2008. *Nature, not human activity, rules the climate: Summary for policymakers of the Report of the Nongovernmental International Panel on Climate Change.* Heartland Institute, Chicago.

SKINNER, C.N. 2007. Silviculture and forest management under a rapidly changing climate. In *Restoring fire-adapted ecosystems: Proceedings of the 2005 national silviculture workshop,* Powers, R.F. (ed.). PSW-GTR-20. USDA Forest Service, Pacific Southwest Research Station, Albany, CA.

SKOG, K.E., AND G.A. NICHOLSON. 1998. Carbon cycling through wood products: The role of wood and paper products in carbon sequestration. *Forest Products Journal* 48(7/8):75–83.

SMITH, J., AND L. HEATH. 2004. Carbon stocks and projections on public forestlands in the United States, 1952–2040. *Environmental Management* 33(4):433–442.

SMITH, J.R., W. RICHARDS, AND E.C. SHEA. 2007. *25x'25 Action plan: Charting America's energy future.* Environmental Action Coalition, Washington, DC.

SOCIETY OF AMERICAN FORESTERS (SAF). 2002. *Wildfire management: A position of the Society of American Foresters*. Available online at www. safnet.org/policyandpress/psst/fire0902.cfm; accessed October 11, 2007.
SOCIETY OF AMERICAN FORESTERS (SAF). 2004. *Preparing a community wildfire protection plan: A handbook for wildland-urban interface communities*. Available online at www.safnet.org/policyandpress/cwpp.cfm; accessed October 15, 2007.
SOCOLOW, R.H., AND S.W. PACALA. 2006. A plan to keep carbon in check. *Scientific American* 295(3):50–57.
SOLOMON, A., R. BIRDSEY, L. JOYCE, AND J. HAYES. 2008. Global change research strategy 2009–2019. Available online at www.fs.fed.us/psw/topics/climate_change/pdf/gc_research_ strat.pdf; accessed March 20, 2008.
SOLOMON, S., D. QIN, M. MANNING, Z. CHEN, M. MARQUIS, K.B. AVERYT, M. TIGNOR, AND H.L.MILLER (EDS.). 2007. *Summary for policymakers. Climate change 2007: The physical science basis. Contribution of Working Group I to the Fourth Assessment Report of the Intergovernmental Panel on Climate Change*. Cambridge University Press, Cambridge, UK.
SPITTLEHOUSE, D.L., AND R.B. STEWART. 2003. Adaptation to climate change in forest management. *BC Journal of Ecosystems and Management* 4(1):1–11.
STAVINS, R.N. 2001. *Experience with marketbased environmental policy instruments*. Discussion paper 01–58. Resources for the Future, Washington, DC.
STENSETH, N.C., A. MYSTERUD, G. OTTERSEN, J.W. HURRELL, K.-S. CHAN, AND M. LIMA. 2002. Ecological effects of climate fluctuations. *Science* 297:1292–1296.
STERN, N., S. PETERS, V. BAKHSHI, A. BOWEN, C. CAMERON, S. CATOVSKY, D. CRANE, S. CRUICKSHANK, S. DIETZ, N. EDMONDSON, S.-L. GARBETT, L. HAMID, G. HOFFMAN, D. INGRAM, B. JONES, N. PATMORE, H. RADCLIFFE, R. SATHIYARAJAH, M. STOCK, C. TAYLOR, T. VERNON, H. WANJIE, AND D. ZENGHELIS (EDS.). 2006. *Stern review on the economics of climate change*. Cambridge University Press, Cambridge, UK.
THE NATURE CONSERVANCY. n.d. *Air quality regulations and smoke management*. Available online at www.tncfire.org/documents/FMR/airquality. pdf; accessed November 17, 2007.
TRENBERTH, K.E., P.D. JONES, P. AMBENJE, R. BOJARIU, D. EASTERLING, A.K. TANK, D. PARKER, F. RAHIMZADEH, J.A. RENWICK, M. RUSTICUCCI, B. SODEN, AND P. ZHAI. 2007. Observations: Surface and atmospheric climate change. In *Climate change 2007: The physical science basis. Contribution of Working Group I to the Fourth Assessment Report of the Intergovernmental Panel on Climate Change*, Solomon, S., D. Qin, M. Manning, Z. Chen, M. Marquis, K.B. Averyt, M. Tignor and H.L. Miller (eds.). Cambridge University Press, Cambridge, UK.
TSS CONSULTANTS. 2006. *Biomass conversion factors and glossary*. Bioenergy and Wood Products Conference, Denver, March 14–16. Available online at www.nationalbiomassconference.org/presentations/BioenergyConv-Glossary.pdf; accessed February 6, 2008.
TYSON, K.S., C.J. RILEY, AND K.K. HUMPHREYS. 1993. *Fuel cycle evaluations of biomass-ethanol and reformulated gasoline*. Technical report 463-4950. National Renewable Energy Laboratory, Golden, CO.
US DEPARTMENT OF ENERGY (DOE). 2006. *Guidelines for voluntary greenhouse gas reporting, final rule. Technical guidelines: Voluntary reporting for greenhouse gases (1605(b)) program*, Part I: Appendix D. US Department of Energy, Office of Policy and International Affairs, Washington, DC.
US DEPARTMENT OF ENERGY (DOE). 2007. *Technical guidelines for voluntary reporting of greenhouse gases program (1605b)*, Chapter 1, Part I, Forestry emissions. Available online at www.pi.energy.gov/enhancingGHGregistry/documents/January2007_1605bTechnicalGuidelines.pdf; accessed November 13, 2007.
US DEPARTMENT OF THE INTERIOR (USDI) AND US FOREST SERVICE (USFS). 2007. *Healthy forests report: FY 2006 final accomplishments*. Available online at www.forestsandrangelands.gov/reports/documents/healthyforests/2006/2006_year_end_report.pdf; accessed November 18, 2007.
US ENVIRONMENTAL PROTECTION AGENCY (EPA). 1990. *The determination of mitigation under the Clean Water Act Section 404(b)(1) guidelines*. Memorandum of agreement between the Department of Army and the Environmental Protection Agency. Available online at www.epa.gov/owow/wetlands/regs/mitigate.html; accessed November 27, 2007.
US ENVIRONMENTAL PROTECTION AGENCY (EPA). 2005. *Greenhouse gas mitigation potential in U.S. forestry and agriculture*. EPA 430-R-05-006. Office of Atmospheric Programs, Washington, DC.

US ENVIRONMENTAL PROTECTION AGENCY (EPA). 2006. *Nitrous oxide.* Available online at www.epa.gov/nitrousoxide/sources.html; accessed December 10, 2007.

US ENVIRONMENTAL PROTECTION AGENCY (EPA). 2007a. *Greenhouse gas impacts of expanded renewable and alternative fuels use.* EPA 420-F-07-035. Office of Transportation and Air Quality, Washington, DC.

US ENVIRONMENTAL PROTECTION AGENCY (EPA). 2007b. *Inventory of U.S. greenhouse gas emissions and sinks: 1990*–2005. EPA 430-R-07-002. Washington, DC.

US ENVIRONMENTAL PROTECTION AGENCY (EPA). 2007c. *Plain English guide to the Clean Air Act: Understanding the Clean Air Act.* Available online at www.epa.gov/air/caa/peg/understand. html; accessed November 17, 2007.

US ENVIRONMENTAL PROTECTION AGENCY (EPA). 2008. *Past climate change.* Available on-line at www.epa.gov/climatechange/science/ pastcc.html; accessed January 4, 2008.

US FOREST SERVICE (USFS). 2004. *Pacific Northwest Research Station science update*, Issue 6, January, Neilson, R., J. Lenihan, R. Drapek, and D. Bachelet (ed.). Available online at www.fs.fed.us/pnw/pubs/science-update-6.pdf; accessed December 10, 2007.

US FOREST SERVICE (USFS). 2005. *CORRIM report on environmental performance measures for renewable building materials.* FS 20050601-1. USDA Forest Service, Forest Products Laboratory, Madison, WI.

US FOREST SERVICE (USFS). 2007. *U.S. Forest Service fiscal year 2008 President's budget overview.* US Department of Agriculture.

UNITED NATIONS FRAMEWORK CONVENTION ON CLIMATE CHANGE (UNFCCC). 2002. *Report of the Conference of the Parties on the Seventh Session, held at Marrakesh, October, 29 to November 10, 2001.* FCCC/CP/2001/13/Add.1. United Nations Framework Convention on Climate Change.

UNITED NATIONS FRAMEWORK CONVENTION ON CLIMATE CHANGE (UNFCCC). 2007a. *Guide to do a CDM project activity—Afforestation and reforestation CDM project activities.* Available online at cdm.unfccc.int/Projects/pac/pac_ar. html; accessed October 12, 2007.

UNITED NATIONS FRAMEWORK CONVENTION ON CLIMATE CHANGE (UNFCCC). 2007b. *The mechanisms under the Kyoto Protocol: The Clean Development Mechanism, Joint Implementation and emissions trading.* Available online at unfccc.int/kyoto_protocol/mechanisms/items/1673.php; accessed October 12, 2007.

UNITED NATIONS FRAMEWORK CONVENTION ON CLIMATE CHANGE (UNFCCC). 2007c. *Status of ratification of the UNFCCC.* Available online at unfccc.int/essential_background/convention/items/2627.php; October 12, 2007.

VALSTA, L., B. LIPPKE, J. PEREZ-GARCIA, K. PINGOUD, J. POHJOLA, AND B. SOLBERG. 2008. Use of forests and wood products to mitigate climate change. In *Managing forest ecosystems: The challenge of climate change,* Bravo, F. (ed.). Springer, New York.

VAN MANTGEM, P.J., AND N.L. STEPHENSON. 2007. Apparent climatically induced increase of tree mortality rates in a temperate forest. *Ecology Letters* 10:909–916.

VIDAL, J. 2007. China could overtake U.S. as biggest emissions culprit by November. *The Guardian,* April 25.

VOLNEY, W.J.A., AND R.A. FLEMING. 2000. Climate change and impacts of boreal forest insects. *Agriculture, Ecosystems and Environment* 82:283–294.

WALTHER, G.-R., E. POST, P. CONVEY, A. MENZEL, C. PARMESAN, T.J.C. BEEBEE, J.-M. FROMENTIN, O. HOEGH-GULDBERG, AND F. BAIRLEIN. 2002. Ecological responses to recent climate change. *Nature* 416:389–395.

WANG, M., Y. WU, AND A. ELGOWAINY. 2005. *Operation manual: GREET Version 1.7, ANL/ ESD/05–03.* Center for Transportation Research, Argonne National Laboratory, Argonne, IL. November.

WATSON, R.T., I.R. NOBLE, B. BOLIN, N.H. RAVINDRANATH, D.J. VERARDO, AND D.J. DOKKEN. 2000. Land use, land use change and forestry. Intergovernmental Panel on Climate Changespecial report. Available online atwww. grida.no/climate/ipcc/land_use; November 17, 2007.

WERNER, B. 2000. *California Air Resources Board memorandum to D. Shimp.* August 17. Available online at www.arb.ca.gov/ei/see/memo_ ag_emission_factors.pdf; accessed January 28, 2008.

WESTERLING, A.L., H.G. HIDALGO, D.R. CAYAN, AND T.W. SWETNAM. 2006. Warming and earlier spring increase western U.S. forest wildfire activity. *Science* 313:940–943.

WESTERN FORESTRY LEADERSHIP COALITION. 2007. *Climate change and western forests: Policy statement.* Available online at www.wflccenter. org/news_pdf/275_pdf.pdf; accessed February 3, 2008.

WESTERN GOVERNORS' ASSOCIATION. 2006. *Clean and Diversified Energy Initiative, Biomass Task Force report.* Available online at www.westgov.org/wga/initiatives/cdeac/Biomass-full.pdf; accessed October 15, 2007.

WIEDINMYER, C., AND J. NEFF. 2007. Estimates of CO_2 from fires in the United States: Implications for carbon management. *Carbon Balance and Management* 2:10.

WILLIAMS, D.W., AND A.M. LIEBHOLD. 2002. Climate change and the outbreak ranges of two North American bark beetles. *Agricultural and Forest Entomology* 4:87–99.

WINISTORFER, P., Z. CHEN, B. LIPPKE, AND N. STEVENS. 2005. Energy consumption and greenhouse gas emissions related to the use, maintenance, and disposal of a residential structure. *Wood and Fiber Science* 37:128–139.

WITTIG, V.E., C.J. BERNACCHI, X.-G. ZHU, C. CALFAPIETRA, R. CEULEMANS, P.DEANGELIS, B. GIELEN, F. MIGLIETTA, P.B. MORGAN, AND S.P. LONG. 2005. Gross primary production is stimulated for three *Populus* species grown under free-air CO_2 enrichment from planting through canopy closure. *Global Change Biology* 11:644–656.

WOODWELL, G.M., F.T. MACKENZIE, R.A. HOUGHTON, M. APPS, E. GORHAM, AND E. DAVIDSON. 1998. Biotic feedbacks in the warming of the Earth. *Climatic Change* 40:495–518.

WORLD RESOURCES INSTITUTE (WRI). 2007. *Climate Analysis Indicators Tool (CAIT) version 4.0.* Available online at cait.wri.org; accessed October 10, 2007.

WULLSCHLEGER, S.D., T.J. TSCHAPLINSKI, AND R.J. NORBY. 2002. Plant water relations at elevated CO_2—Implications for water-limited environments. *Plant, Cell and Environment* 25:319–331.

ZERBE, J.I. 1983. *Energy properties of wood.* USDA Forest Service, Forest Products Laboratory, Madison, WI.

ZERBE, J.I. 2006. Thermal energy, electricity, and transportation fuels from wood. *Forest Products Journal* 56(1):6–14.

ZHOU, G., S. LIU, Z. LI, D. ZHANG, X. TANG, C. ZHOU, J. YAN, AND J. MO. 2006. Old-growth forests can accumulate carbon in soils. *Science* 314(5804):1417.

ZVEREVA, E.L., AND M.V. KOZLOV. 2006. Consequences of simultaneous elevation of carbon dioxide and temperature for plant–herbivore interactions: A metaanalysis. *Global Change Biology* 12:27–41.

Subject Index

About the Authors

Robert W. Malmsheimer *Associate Professor of Forest Policy and Law, SUNY College of Environmental Science and Forestry, Syracuse, New York*

Malmsheimer has been a professor at SUNY ESF since 1999 and teaches courses in natural resources policy and environmental and natural resources law. His research focuses on how laws and the legal system affect forest and natural resources management, including how climate change and carbon sequestration policies affect forest and natural resources. Prior to becoming a professor, Malmsheimer practiced law for six years. He has a Ph.D. in forest policy from SUNY ESF, a J.D. from Albany Law School, and a B.L.A. from SUNY ESF. He was the 2007 chair of the SAF Committee on Forest Policy and served on the committee from 2005 to 2007. He has served on numerous national and state SAF committees and task forces.

Patrick Heffernan *President, PAFTI, Inc., Hungry Horse, Montana*

Heffernan began his career in forestry in 1976 and graduated with a national diploma in forestry from the Cumbria College of Agriculture and Forestry in 1981. He has enjoyed a variety of forestry employment in Great Britain and the United States and is now part owner and manager of an experimental private forest in New Zealand. He has been an SAF member since 1990, served as chapter and state chairs in Montana, and is currently on SAF's National Policy Committee. He was involved during the formative years of what has become the National Carbon Offset Coalition, where his interest in promoting inclusive solutions to forest carbon sequestration posited an approach for scientific net primary productivity calculations as a sound basis for forestry carbon credit markets.

***Steve Brink** Vice President–Public Resources, California Forestry Association, Sacramento, California*

Brink has been with the California Forestry Association since July 2005. He represents most of the remaining solid wood mill infrastructure and many of the remaining biomass powerplants in the state. His focus is on timber and biomass wood supply from the national forests, which manage 50 percent of the state's productive forestland. Since 2007, Brink has focused on forest carbon sequestration, carbon life-cycle modeling, forestry protocols, and the potential of renewable energy credits for forest landowners, wood manufacturing facilities, and biomass for power generation. He graduated from the University of California at Davis with a degree in civil engineering. Prior to joining the California Forestry Association, he spent 36 years with the US Forest Service.

***Douglas Crandall** Director of Legislative Affairs, US Forest Service, Washington, DC*

Crandall is currently director of Legislative Affairs for the US Forest Service. Previously, for eight years, he was the staff director for the US House of Representatives Subcommittee on Forests and Forest Health, with jurisdiction over most legislation and oversight concerning the Forest Service and Bureau of Land Management. He also served with the Society of American Foresters as policy director, the National Forest Foundation as vice president, and the American Forest and Paper Association as director responsible for national forest issues. Earlier in his career, he spent 10 years managing a lumber company in Livingston, Montana, and four years on the Brazilian Amazon, first as a forester and float-plane pilot, then as a plywood mill manager. Doug graduated with a B.S. in forestry from Oregon State University. He has been a member and officer of numerous forestry, industry, conservation, and community organizations.

***Fred Deneke** Staff Advisor, 25x25 Renewable Energy Alliance, Prescott, Arizona*

Deneke is forestry staff advisor to the 25x25 Renewable Energy Alliance on woody biomass and biomass energy. He also administers a woody biomass cooperative agreement involving the National Association of Conservation Districts, the US Forest Service, and the Bureau of Land Management. Deneke retired from the US Department of Agriculture in 2005 after 32 years of service with the US Forest Service and USDA Extension Service. Prior to his retirement he was assistant director for the Cooperative Forestry Staff of theUS Forest Service in Washington, DC, where he served as national lead for woody biomass utilization for State and Private Forestry on the US Forest Service Woody Biomass Team and the served on the Interagency Woody Biomass Utilization Group (involving the Forest Service, Department of Interior, Department of Energy, other federal agencies, and nongovernmental partners). Prior to his USDA service, Deneke was an assistant professor at Kansas State University. He is a graduate of Colorado State University with a B.S. in forestry science. He also has M.S. and Ph.D. degrees in horticulture from Kansas State University.

Christopher Galik *Research Coordinator for the Climate Change Policy Partnership, Duke University, Durham, North Carolina*

Galik currently serves as research coordinator for the Climate Change Policy Partnership, a collaborative project intended to leverage the resources of Duke University to determine practical strategies to respond to climate change. Within this partnership, Galik has primary oversight over biological carbon sequestration, biomass, and biofuel research activities. Previously, he spent several years in Washington, DC, as a policy analyst, specializing in a variety of environmental issues, including species conservation and federal forest management and policy. He holds a master of environmental management degree from the Nicholas School of the Environment at Duke, with a concentration in forest resource economics and policy. He received his B.A. in biology from Vassar College. Galik has been an SAF member since 2001.

Edmund A. Gee *Woody Biomass Utilization Team Leader, US Forest Service, Forest Management, National Forest System, Washington, DC*

Gee is the Chair, Interagency Woody Biomass Utilization Group and the National Woody Biomass Utilization Team leader for the US Forest Service. He oversees the Woody Biomass Utilization Group and Team in the development of sustainable woody biomass strategic planning, policy, and implementation of the plan as it relates to climate change. Gee works directly with the Chief's Office and the Washington Staff Directors Woody Biomass Steering Committee to work across all deputy chief areas as well as with the Departments of Interior, Energy, and Defense, the Environmental

Protection Agency, and other USDA agencies. He received his B.S. degree in natural resource management from the University of California at Berkeley, a Certified Silviculturist from Oregon State University and University of Washington, and an M.B.A. from the University of Phoenix. He has been a member of SAF since 1983.

John A. Helms *Professor Emeritus, University of California, Berkeley, California*
Helms joined the faculty of the School of Forestry, Berkeley, in 1964 and has M.S. and Ph.D. degrees from the University of Washington, Seattle. At Berkeley he was professor of silviculture and became head of the department. Much of his research was in tree physiology with emphasis on net uptake of carbon dioxide by mature trees in relation to stresses from water availability, temperature, and air pollution.He served SAF as chair of the Forest Science and Technology Board for two terms and as president in 2005. He gave testimony twice before Congress in 2007 on climate change effects on forests and wildfires. He currently serves on the Board of the California Forest Products Commission, in 2007 was appointed a member of the Sustainable Forest Initiative's External Review Panel, and in 2008 to the US/Canada Forest Carbon Standards Committee.

Nathan McClure *Director, Georgia Forestry Commission's Forest Products Utilization, Marketing, and Development Program, Macon, Georgia*
McClure currently leads the Georgia Forestry Commission's Forest Products Utilization, Marketing, and Development program and also serves as the director of Forest Energy for the agency. He has worked in a variety of positions over the past 24 years with the commission. McClure is a Georgia Registered Forester and a SAF Certified Forester. He received the SAF Presidential Field Forester Award in 2005. He is a 1983 graduate of the University of Georgia with a B.S. in forest resources management. Recent assignments include creating additional values from Georgia's forests through marketing and new product development, facilitating the development of a forest biomass energy industry, and initiating Georgia's new carbon sequestration registry, as well as working with traditional forest products industries.

Michael Mortimer *Director of Natural Resources Program, Virginia Tech College of Natural Resources, National Capitol Region, Falls Church, Virginia*

Mortimer is currently the Director of the Natural Resources graduate program for Virginia Tech's College of Natural Resources campus serving the Washington, DC area. Previously he acted as the director of forest policy for the Society of American Foresters, and served on the faculty of the Virginia Tech Department of Forestry, where he carried out research and teaching in the areas of public and private land forest management and regulation and published in the areas of forest and biodiversity conservation. He also served as an assistant attorney general for the Montana Department of Natural Resources and Conservation, where he advised and litigated on behalf of the agency's forestry programs. Mortimer received his Ph.D. in forestry from the University of Montana, his J.D. from the Pennsylvania State University, and a B.A. from Washington and Jefferson College. He has been an active member and officer of various professional, forestry, and policy-related organizations at both national and state levels.

Steve Ruddell *Senior Program Officer, Forest Carbon Portfolio, World Wildlife Fund, Washington, D.C.*

Within the Forest Carbon Network Initiative for the World Wildlife Fund, Ruddell manages forest carbon project development that reduce emissions from deforestation and degradation (REDD) in the world's major tropical forest regions, and accesses voluntary and compliance markets for these projects. Ruddell previously consulted with clients on investments in forest conservation and sustainability initiatives using market-based mechanisms, including carbon asset management strategies for trading forest carbon offset projects in voluntary forest carbon markets. Ruddell also conducted Chicago Climate Exchange (CCX) and Over-the-Counter forest carbon asset management services in North and South America, including forest offset project economic analyses, development, quantification, verification. He is on the revision committee for the second edition of the Community, Climate, and Biodiversity standards, Chair of the SAF's Working Group on Bio-energy, Climate Change, and Carbon; Vice Chair for the Bi-national Forest Carbon Standards Committee, and a member of the CCX Avoided Deforestation Committee. Ruddell received his B.S. in forest management from Utah State University, and an M.S. in forestry and an M.B.A. in operations management from Michigan State University, and has conducted post graduate work in Resource Economics.

***Matthew Smith** Director of Ecosystem Services, Forecon EcoMarket Solutions LLC., Falconer, New York*

Smith, ACF, CF, EMS-A, is the director of Ecosystem Services for Forecon EcoMarket Solutions LLC (Forecon EMS), an approved aggregator for the Chicago Climate Exchange (CCX), and director of Land Management for Forecon, Inc., a forestry consulting company and CCX verifier. Since the mid-1990s he has managed the Land Management Department for Forecon, Inc., working with large land management clients such as timber investment management organizations and high net worth individuals. In 2004 Smith began working in the area of ecosystem services on behalf of these clients and has conducted forest carbon modeling and analysis, prepared managed forest carbon offset projects for the CCX market, and written about carbon sequestration for national audiences. He has served as a consultant to state and local municipalities, forest owner organizations, carbon registries, professional organizations, private landowners, CCX, and other groups. He also directs a team of ecosystem specialists working on marketbased incentives for biodiversity and water resources. Smith serves as the chair of the Western New York Chapter of SAF and is a member of the Forest Carbon Education Group, the 25x25 Carbon Working Group, the Association of Consulting Foresters, and the New York Forest Owners Association. He holds a B.S. in forest resource management from the SUNY College of Environmental Science and Forestry and is an auditor and consultant for sustainable forest certification systems.

***John C. Stewart** Biomass and Forest Health Program Manager, Department of Interior, Office of the Secretary's Office of Wildland Fire Coordination, Washington, DC*

Stewart, the Biomass and Forest Health Program manager for the Department of Interior, represents the department on biomass utilization for renewable energy under the National Energy Plan and leads its efforts at small wood utilization under the National Fire Plan. Stewart also led an interdepartmental team in writing a joint woody biomass policy for the Departments of Interior, Energy, and Agriculture. Previously, Stewart was a forester with the Bureau of Land Management in Washington, DC. He had 22 years of experience with the US Forest Service throughout California before joining BLM. Stewart received a B.S. degree from the University of California at Berkeley and also worked for Dr. Ed Stone doing basic research in seedling growth response and vegetation descriptions. He has been a member of SAF since 1978 and served as the Bay Area Chapter chair in 1990 and 1991.